Comparative Marine Policy

Perspectives from Europe, Scandinavia, Canada and the United States

Center for Ocean Management Studies
University of Rhode Island

PRAEGER

PRAEGER SPECIAL STUDIES • PRAEGER SCIENTIFIC
A J.F. BERGIN PUBLISHERS BOOK

Library of Congress Cataloguing in Publication Data

Rhode Island. University. Center for Ocean
Management Studies.
Comparative marine policy.

Bibliography: p.
1. Marine resources and state. 2. Marine
resources conservation. 3. Coastal zone
management. 4. Fishery management. I. Title.
GC1017.R46 1980 33.9'164 80-21455
ISBN 0-03-058307-1

Published in 1981 by Praeger Publishers
CBS Educational and Professional Publishing
A Division of CBS, Inc.
521 Fifth Avenue, New York, New York 10017, USA

J.F. Bergin Publishers, Inc.
670 Amherst Road
South Hadley, Massachusetts 01075

Second Printing

0123456789 056 987654321

Printed in the United States of America

Preface

There have been few systematic, empirically based, country-by-country comparative analyses of marine policy. Most of the current studies deal with international relations; however, researchers in Europe, Scandinavia, Canada, and the United States have conducted and are continuing to conduct sophisticated analyses of the particular policy needs of their countries. In order to provide an opportunity to share insights and compare the marine policy, structures and processes of these countries, the Center for Ocean Management Studies at the University of Rhode Island convened an international conference on **Comparative Marine Policy: Perspectives from Europe, Scandinavia, Canada, and the United States.** This volume is a compilation of the invited papers and a summary of the conference discussions.

The chapters in this book generally reflect the conference program. Chapters 1 and 2 concern the opening session of the conference, which set the tone for subsequent discussions. In this session Dr. Edward Wenk from the United States summarized in his keynote address the Global Principles for National Marine Policies. Later Dr. Stjepan Keckes from the United Nations discussed the concept of Regional Seas: An Emerging Marine Policy Approach. Other speakers from the United States, Canada and Norway discussed the State of the Art in Comparative Marine Policy. The remaining chapters represent subsequent sessions in which speakers from eight countries addressed specific policy issues, such as coastal management, offshore oil and gas policy, fisheries management, and marine environmental protection. The last chapter summarizes the concluding session of the conference which focused on New Directions in Comparative Marine Policy.

The conference was sponsored by the Center for Ocean Management Studies (COMS) at the University of Rhode Island (URI) in cooperation with the National Oceanic and Atmospheric Administration (NOAA). The Center for Ocean Management Studies was created in the fall of 1976 for the purpose of promoting effective coastal and ocean management. This is achieved by providing a constructive forum for communication, research, and education on marine management issues. We feel that this volume represents such a forum and provides valuable insights into national efforts to manage the marine environment. It is hoped that all countries will benefit from sharing these experiences and more effective marine policies will result.

Acknowledgements

Many people contributed to this effort. First and foremost, I would like to thank the conference chairman and members of the program committee who contributed ideas and served as session chairmen or participants:

Conference Chairman:

Timothy M. Hennessey, Political Science, URI

Program Committee:

Dennis Callaghan, Organizational Management, URI
Francis X. Cameron, Geography and Marine Affairs, URI
Thomas Grigalunas, Resource Economics, URI
Eric D. Schneider, Center for Ocean Management Studies, URI, and US Environmental Protection Agency
Lars Vidaeus, New England Regional Fishery Management Council (presently with World Bank)

I would also like to acknowledge the members of the international advisory committee who provided invaluable assistance in identifying appropriate speakers:

International Advisory Committee:

Barry Buzan, University of Warwick, England
Robert Friedheim, University of Southern California, United States
Michael Glazer, National Oceanic and Atmospheric Administration, United States
Edgar Gold, Dalhousie University, Canada
Marc Hershman, University of Washington, United States
Douglas Johnston, Dalhousie University, Canada
Edward Miles, University of Washington, United States
James Rickard, Exxon Production Research Company, United States
Arild Underdal, Nansen Foundation, Norway

In addition, a special thanks should be given to the conference speakers who made this volume possible.

Numerous individuals and organizations provided logistical and technical support in organizing the conference and preparing the proceedings. However, special recognition should go to Carol Dryfoos, COMS Administrative Assistant, Nancy Ingham, COMS Technical Editor, the URI Conference Office, and J.F. Bergin Publishers, Inc. Last but not least a special thanks is extended to the National Oceanic and Atmospheric Administration: Coastal Zone Management, National Marine Fisheries Service, Policy and Planning, and Sea Grant, for funding the conference.

Virginia K. Tippie
Executive Director
Center for Ocean Management Studies

Introduction

This book proceeds from the assumption that considerable benefits can be derived by marine specialists from advanced industrial democracies through the sharing of information on fisheries, coastal zone management, offshore oil and gas, and environmental protection.

Each of the countries represented in this volume has attempted to establish regulating regimes and processes as these apply to the resources in question. The manner in which they have done so and the differences in outcomes given variations in institutional design, structure, and process could constitute the subject matter for comparative marine-policy studies. Indeed, it is hoped that the discussions and papers represented here can establish the basis for a series of studies which will investigate regulatory regimes as these apply to marine policy in advanced industrial democracies.

Some of the basic questions which might be posed in this regard are:

1. Does the cost of regulation exceed the social benefits?
2. Are centralized or decentralized regulatory regimes to be preferred, and why?
3. How do regulatory structures and outcomes vary across different resources, and why?

It is my opinion that these questions can be answered only through a concerted, interdisciplinary, cross-national approach. Moreover, the probabilities of success will be enhanced considerably by a close cooperation between government decision-makers and industry and academic analysts.

It is hoped that this volume will contribute in some measure to the development of comparative marine studies that will enhance our ability to manage the valuable resources in question.

Timothy M. Hennessey
Conference Chairman

Contents

ONE

Global Principles For National Marine Policy

CHAPTER 1

Global Principles for National Marine Policies: A Challenge for the Future

EDWARD WENK, JR.
University of Washington
United States

BACKGROUND: THE GLOBAL SITUATION

Geopolitical dynamics of the sea have been recognized, implicitly if not explicitly, ever since man set out in log canoes to explore lands lying beyond the horizon, and to claim those lands as extensions of national territories. Then, the oceans afforded routes for trade and occupation that progressively became more firmly developed and defended. These same oceans later entered military strategy, serving both as protective moats and as levers in building empires. Nations soon drew boundaries of maritime property. By and large, people recognized the sea as a hazardous and hostile environment, to be suffered through in transit to the goals and treasures that lay beyond. Only recently has the full resource potential of the sea been linked to human destiny.

It was probably the community of marine scientists that first recognized the global dimensions of the sea, and the necessity for their study on a politically neutral, unified basis. Indeed, cooperation among scientists to develop better understanding, then to propose rational controls over exploitation of marine resources, dates back almost one hundred years. The scientific community still represents the most vigorous advocate of multinational cooperation and of freedom in the collection of data and their dissemination.

Within the past thirty years, however, a new economic, social, and political latticework criss-crossed the sea, binding people, nations and the oceans together in one functional—if not one political—world. Now commerce, communications, individuals, knowledge, ideas, culture, and even pollutants traverse boundaries, no longer constrained either by political geography or by topography. Yet, individual nations continue to generate domestic marine policy as though its implementation could be achieved without connections to the rest of the world. While this tunnel vision may be attributed either to a renaissance of nationalistic self-interest or to lack of imagination, it is seriously anachronistic. Indeed, such provincial views could well lead to policy assumptions, policy design or implementation strategies that in the long run may even be counter-productive to a nation's own self-interest. Obsolete policies may then become part of the problem, rather than part of the solution.

PROVINCIALISM IN POLICY DESIGN

In short, there is no longer an impermeable—much less definable—boundary between domestic and foreign policy for any nation in its relation to the sea.

That parochialism has a major corollary. There is a further tendency for nations to develop their domestic marine policies as though they were separate from other domestic policies. In the first instance, marine policy often deals with means rather than with ends. But the relevant ends concern national security policy, inflation, employment and economic policy, energy policy, environmental policy. What this means in operational terms, therefore, is that even when dealing with the internal content of separate and often piecemeal elements of marine policy, there needs to be a sensitive appreciation of their interconnections with broader policy arenas which have reason to claim strongest attention both from political leadership and from citizens.

Finally, there is a third mode of provincialism, one I would call *temporal* provincialism. For in the development of marine policy, as with other policies, one finds a melancholy imbalance between the short and the long run. That is to say, motivation of national policies has been goaded by crisis, by pressure from militant special interests, or by anticipations of short-term benefits. In the political bargaining process no one represents the future; thus, as compromise is reached, it is usually between contending parties that have been arguing from their individual short-term perspectives.

This myopic property of decision making is loaded with traps. First of all, many of the policy initiatives which appear to have clear benefits in the short run are burdened with lateral and longitudinal impacts which are so hard to reverse that they entail high economic, ecological, political or social costs. These unwanted penalties are expensive, whether accepted as is or met with measures to mitigate the undesirable effects. Put another way, in the blindness associated with such policies, bills may eventually come due that render the policy obsolete or bankrupt.

Apart from the failure to sense and prevent such externalities, there is another feature of shortsightedness that undermines policy performance. Every policy is predicated on assumptions regarding the state of society. If these assumptions rigidly reject the possibilities of future change, the policies may fail to hit the target. Those who hunt ducks are well aware of the importance of lead—you never hit if you aim directly. You have to point the gun where you think the duck will be when the shot arrives.

Paradoxically, while marine-policy planning may be mired in generally unproductive processes, the demands for new policy are growing more urgent. That is to say, the oceans are increasing as a source of food, minerals and energy, as a transport medium for global trade, as a stimulus for employment and economic vitality, and as a locale for esthetic joys and recreation. Oceans continue to be vital factors in foreign relations and as fundamental elements in nuclear deterrent strategies of the super powers. We should thus expect more—and more complex—sea policies to be generated by all nations concerned. If, however, we continue the past practice of unwittingly constraining marine policies within national boundaries; if we continue to assume that domestic policies may be separated from foreign policy; if we neglect the connections between marine policy and other policies which represent the social goals of each nation; and if we continue to discount the future, we can only look forward—at the very least—to confusion, and more likely to world conflict as nations vie for unilateral influence over their own narrowly-perceived situations and their fates. Then, in belated attempts to create their own futures, they will discover that no nation can control its destiny single-handed.

There are clear lessons here for those engaged in the study and development of marine policy. First of all, it suggests the need to adopt a different attitude and breadth of vision, both geographically and temporally. And it should also prompt those engaged in policy planning—and particularly those with responsibilities and opportunities of political leadership—to recognize that, as each nation contemplates its future, it must take into account the need for a far more holistic and future-oriented approach; one consistent with the practical dynamics of an interdependent technological world as well as committed to the most fundamental human values.

GLOBAL PRINCIPLES FOR NATIONAL POLICIES

One thinks immediately of the need for common principles to guide initiatives by individual nations and lead to agreements among them. One set of principles was recently generated by the Cousteau Society, through a study group that included Jacques Cousteau, the Society's founder, and Messrs. Abel, Adams and Hargis, well-known in the American policy community. I was privileged to serve as chairman. Those proposals follow:

1. Ocean Policies Must Be Global
Because the oceans of the world are essential to life on this planet, because ideas, culture, knowledge, communications, commerce, travel of individuals and of pollutants are less constrained by geographic and political boundaries than ever before in history, and because both use and abuse of the seas are of consequence to all peoples, a *global ocean policy* must be established that defines a common set of principles and rules as guidelines for activities of individual nations and for agreements among nations.

2. We Cannot Shirk our Obligations to Future Generations
Because threats to survival appear to be mounting on a planetary scale, humankind everywhere now shares a common destiny. Moreover, because ocean resources are increasingly recognized as contributing to human needs for food, energy, minerals and for serenity, pressures are growing for swift and unwitting exploitation. To fulfill a moral obligation that the legacy of the oceans that we enjoy be continued, our first concern must be directed to future generations. Thus, all activities involving the oceans should be undertaken only after various options are developed and consequences and risks for our progeny weighed against anticipated short-term, provincial benefits.

3. Ocean Policy Must also Encompass Fresh Water Systems, Polar Areas and the Atmosphere
Lakes, marshes and rivers are the roots of the ocean, carrying to the open sea nutrients and pollutants affecting the productivity and the quality of sea water. Glaciers and polar ice caps have locked in great quantities of water that govern the level of the ocean and are sensitive to natural or man-made climatic changes. Thus, the management of lakes, marshes, rivers, glaciers, and indeed of polar areas and the atmosphere, must be guided by the same principles as those defined by the global ocean policy.

4. Exploitation of Marine Resources Requires Careful Management to Avoid Depletion
To obtain maximum long-term benefits from the oceans,

public and private enterprise must be encouraged to determine the full resource potential of the sea as well as the most effective means for extraction of food, minerals and energy without unwitting depletion of the basic capital of a planetary inheritance. Resource management should go beyond yardsticks of economic efficiency to consider social and environmental impact and interdependencies of ocean policy with those policies concerning the land and human activities generally.

Industrial fishery practices must be modified to conserve and even enhance natural stocks, and aquaculture must be focused on the production of low-cost protein food systems to help meet world nutrition needs. Populations of marine mammals should be maintained at their maximum-yield level.

Deep-sea mineral resources must be mined and processed at sea only if permanent damage to the environment is avoided and if profits and risks are shared by all nations according to internationally established guidelines.

Since the ocean is the most efficient natural concentrator of solar energy, all plausible methods of tapping this inexhaustible renewable energy source must be examined and promising avenues actively developed.

5. Vitality of the Oceans Must Be Protected

Attention must focus on the three most critical zones of the ocean: the coastal margin, the surface, and the bottom waters, especially to guard against mechanical destruction and pollution.

Substantial sections of the world's coastlines should be kept free from any development. Liquid and gaseous effluents from industrial and urban communities must be limited by innovative technology or treated at each source, especially to exclude toxic non-degradable substances. Dumping at sea of harmful waste must be prevented.

6. Knowledge About the Oceans Should Be Sharply Enhanced and Disseminated

To improve the effectiveness and the safety of managing the global ocean and associated water systems:

All states must collaborate in scientific research and exploration, to collect information about the sea, its resources and the effects of human use. The scientific community must respect rights of coastal states, and share results promptly with all interested parties. At the same time, no arbitrary limits should be imposed upon freedom of research. Indeed, knowledge has to become a common heritage of mankind.

Worldwide, education should be broadened and deepened concerning marine ecology and the relationship of the oceans to human affairs. All media should be utilized to enhance understanding by the public at large as well as on scholarly levels.

Conditions of the world water systems and atmosphere must be kept under surveillance, by such universal systems as remote-sensing by satellites. The resulting sustained monitoring of the oceans and regular reports on the state of its health must be part of the cooperative effort of all nations.

7. Action to Improve Safety of Marine Transportation isMandatory

Safety of marine transportation should be sharply improved by adoption and application of international rules in ship design and operation, on personnel qualifications, traffic management and ship/cargo inspections to assure maximum protection to life, property and the marine environment.

8. We Must Strive for Peaceful Use of the Seas

The oceans must be employed for peaceful and cooperative purposes, to advance world understanding, and to reduce conflict and restrain nationalistic territorial ambitions. Deployment of nuclear weapon systems on and under the sea must be ultimately eliminated.

9. Coastal Nations Should Exercise National Responsibilities for International Stewardship

The 200-mile zones under consideration by the Law of the Sea Conference must be regarded by states in the first instance as "zones of responsibility." In these zones, coastal nations must accept and conscientiously implement international rules and standards for prudent stewardship of fisheries, coastal resources and for protection of the environment.

Coastal nations should alert neighboring states in event of a marine environmental accident, and provide assistance to other states to mitigate effects of accidents.

10. Paramount in Promoting Global Policy is the Establishment of a World Ocean Authority

A world ocean authority must be created to integrate presently fragmented international activities. The authority must work toward scientific understanding of the sea as a rational basis for resource management, define and update rules and standards that will assure balanced and equitable exploitation and conservation of marine resources with due regard for obligations to future generations.

The world ocean authority, working with strengthened international institutions, should assist developing nations in strengthening their domestic technical institutions and their staff, so they may share responsibilities as one of a family of nations collectively dedicated to employing the sea to serve humankind.

POLICY PROCESS: THE ROLE OF GOVERNMENT

The next problem we are obliged to deal with here concerns the processes of marine policy-making, as well as the substance. Almost every nation lacks internal coherence among different elements of its marine policy. Yet public policy as a steering mechanism is at the heart of the matter. To be sure, in capitalist countries virtually all activities of the ocean anticipate a clear role for private capital and entrepreneurship. This applies to fishing, to transportation, and to energy and mineral extraction. Nevertheless, even here, the government has been obliged to set the course and the commitment that relates the oceans to our national interests for a number of reasons:

1. Ocean policy seldom reflects a need in itself, but rather serves as a means toward achieving broader purposes, such as national security, food, energy and resource sufficiency, economic stability, protection of public health and safety. Ocean policy must thus be harmonized with other major initiatives directed at contemporary social policies which themselves are a province of national governments.

2. By law both the living and non-living resources of the sea are a common property for which the government acts as steward to assure equitable distribution of benefits from development, the conservation of fishery stocks, prudent rates of withdrawal of non-renewable resources and protection of the environment.

3. Coastal activities along the marine boundaries of different provinces are so ubiquitous that few can be conducted exclusively within the purview of any single unit and thus must be articulated through some overarching medium.

4. Marine research and education have been traditionally supported by government rather than industry because marine-related industries

have been so fragmented and weak, and because observations need to be shared without proprietary protection.

And finally 5. Marine activities manifest so many global interactions that in inter-governmental negotiation the national interest must be represented through a single voice at the federal level.

In the United States this enlightened and vigorous approach to ocean affairs was the purpose of the *Marine Resources and Engineering Development Act* (PL 89-454) of 1966. Its mandate set the stage in relating oceans to national needs. Also, by recognizing so many agencies with fragmented responsibilities in ocean affairs, the President was given statutory responsibility for direction of the entire enterprise and for coordination of the federal participants.

In the four years following the 1966 mandate, a wide range of presidentially-supported initiatives strengthened US marine policy and practice. In more recent years, however, that attention has lagged. Thus, cycles of benign neglect that seem endemic with ocean policies are not automatically arrested by legislation. In the United States, for example, continuous oversight and recharging are clearly needed, by a continuing commitment by the Congress in fulfilling its constitutional role.

In emphasizing this role of government in marine policy, we are obliged then to examine the general setting for government decision-making, especially as it responds to priority concerns, mood, and political styles of each nation.

POLICY PROCESS: SOCIAL AND CULTURAL INFLUENCES

We find in the United States again that citizens no longer support science and technology with the same uncritical enthusiasm that they did for two decades after World War II, and especially for a few years after the Soviet space surprise. Secondly, designing policy has become ever more perplexing and strenuous, partly because technology-related issues are growing more complex and involve new networks of institutions and communities engaged in technological enterprise. Thirdly, special interest groups more zealously lobby their narrow, single-issue objectives and parochial institutional ambitions. Then they collide with one another, so that legitimate differences in viewpoints end in heated and unresolved conflict. Many program and policy advocates focus only on the short run, indifferent to unwanted side effects and to the need to balance short with longer-run considerations. While the government seems to end up as umpire among the contenders, there is then diminished attention on—if not complete neglect of—the public interest which the government is fundamentally presumed to defend. Moreover, when policies are developed piecemeal, when varied federal technological efforts are uncoordinated, when the public itself is uninformed and confused, when it is under the stress of inflation so as to demand crisper government administration with less clumsy regulation at lower cost, it is easy to see that policy making in marine affairs cannot be isolated from the social turmoil and frustration which attend policy making in general.

POLICY PROCESS: NEW CHALLENGES

What may be at stake in policy process, therefore, is innovation in two different but compatible directions. The first is to increase policy-planning capabilities at the highest possible level, so as to facilitate integration of different policies and cross-connections among different agencies responsible for separate implementation. Otherwise, policies developed by single-mission agencies are bound to be parochial, very likely driven by the desire to patrol territory and continue practice of past missions that have not been reviewed and refreshed.

The second consideration relates to incorporation within the policy-planning technique of a much more explicit consideration of the future. Indeed, this is a major defect in policy-making at all levels and in all areas. What is so often forgotten concerns a fundamental characteristic of public policy itself—that it is by nature a bridge between the present and the future. The reason is very simple. When one considers the various time-consuming steps between problem identification, policy design, generation of alternatives, negotiation, decision, then steps of authorization and appropriation, and finally implementation, one should not be surprised to discover a time interval stretching from five to fifteen years. Circumstances existing at the time the policy issue was first identified have now—in a rapidly changing technological culture—undergone major revision, and the policy no longer hits its target.

RELATING GLOBAL PRINCIPLES TO FUTURE-ORIENTED PROCESSES

Incorporating the future into policy design, however, has a far deeper philosophical rather than methodological implication. We must now interlock the substance of marine policy with processes of its generation. Global principles for national marine policies set forth earlier can be put entirely in the context of a framework for public order of the oceans, but they also stem from consideration of marine resources as a public trust. This opens up completely new vistas for defining a *common heritage* of mankind, that brilliant, ennobling concept which has depreciated to hollow rhetoric.

As all peoples on the planet approach the end of the century, a vigorous examination is being undertaken—implicitly if not explicitly—of the human condition. What we find are a number of questions that challenge our very existence. Indeed, current predicaments are sufficiently intense and so potentially lethal that we are obliged to ask some basic questions concerning human survival. (Here, I define survival as being both alive and free.) Where indeed are we? Where are we headed and where should we be headed? Finally, what should we do to avoid some of the perils that are flickering over the horizon?

If it would appear that this theme simply reflects the anxious warnings of another Jeremiah, let us recall how observers of our present situation characterize the threat horizon. We are confronted with:

- the threat of nuclear war or of nuclear destruction brought about through terrorist activity, blackmail or theft;
- the danger of widespread famine, starvation and resulting world disorder;
- the danger of environmental poisoning;
- the danger of inadvertent climate modification;
- the dangers of large-scale global disorder arising from economic disparity among nations, now stressed by the burden of increased energy costs;
- the dangers of failures in our social institutions to match the pulses of innovation and change engendered by a technological culture;
- the dangers of pathological shifts in values that threaten freedoms even in those parts of the world that are constitutional democracies:

Although the human species has survived through continuous exposure to peril over tens of thousands of years, our sense of reality tells us that the number of simultaneous threats, their scale in terms of potentially affected populations, along with new orders of innocent exposure, may have exhausted most of our resilience against ecological insult or psychological deprivation. They are altogether without precedent. Survival, not property, is our common heritage.

COLLECTIVE AS WELL AS NATIONAL SECURITY

We must shift our attention beyond notions of national security to the additional consideration of collective security. In so doing, we must recognize that most of these threats have been enhanced by technology. Nevertheless, technology is so firmly engraved on our culture that there is no turning back. One reason for that affectionate engagement is that we have found life to be ennobled by technology. It has afforded for hundreds of millions a new scale of human freedoms—freedom from poverty and despair, from backbreaking labor, from disability and disease, from ignorance, from limitations of geography, of environment and of social deprivation. Technology must be the major engine of development to free the Third World from its historic deprivation.

The unnerving counterpoint to this benificence is found in the unwanted side effects—social, economic, environmental and political impacts—that pose jeopardy of servitude or extermination. The blame, however, cannot be placed on a mindless technology. We must recognize that today the key decisions regarding technology concern the selection of goals to which it is directed, the resources invested, the strategies of implementation, indeed, the stark questions of who wins, who loses and how much. Most of these decisions are made by national governments, so the challenge is how to tune public policy concerned with technology so as to produce socially satisfactory outcomes. What *is* socially satisfactory cannot be evaluated entirely in terms of costs and benefits to populations contained within historic political boundaries. In underscoring the interconnectedness of functions on the planet previously outlined, we must recognize a situation where events anywhere have effects everywhere. All nations are locked together. Like it or not, we now face a common fate.

Since technology concentrates power and wealth and plays a political role in our society, we are obliged to consider with far greater sensitivity and perception than ever before the design of policies—in this case marine policies—whose geographical dispersal and future effects belie seemingly bounded initiatives.

A DOCTRINE OF ANTICIPATION

Upgrading and updating techniques in policy analysis to include such planetary and future considerations fundamentally entails a *doctrine of anticipation.* Put another way, it is essential to consider what the foreseeable trends may suggest as our destiny, and what plausible alternatives may be considered that would gain the intended benefits with a minimum of unwanted side effects. This is a commonly understood practice of looking before you leap, of building in the concept of early warning. This is not idealism. It is a pragmatic response to common dangers.

At the same time, our sense of reality tells us that all governments—capitalist, socialist and developing—have an increased propensity to deal with the short run at the expense of the long. All decision-making theatres seem to be deaf to signals about the future. Indeed, all governments and all peoples are suffering from pathologies of the short run, which inadvertently blind us to the unwanted side and future effects of our decisions. Worse, the process of neglecting the future may actually be undermining the decision mechanism itself. These become the political limits in steering technology.*

We have to accept the unnerving phenomenon that decision-makers everywhere are perplexed by new levels of complexity and hyperinterdependence in our society, by uncertainty, by a heightened pace of social change, by the inutility of past experience, the hauling and pulling in the political theatre as each narrow interest advocates its own short-range goals. Moreover, there are distortions in public communication, gaps in understanding, inflamed institutional tribalism and a paradox that in a richer world resources less and less match rising expectations. The competition for them thus becomes more strenuous. No small wonder that we appear to just muddle through, steering by a form of political dead reckoning, operating with tactics of opportunistic incrementalism. To a marine audience, the metaphor is something like navigating a ship at night on stormy seas without charts and without even knowing whose hand is on the rudder.

Under these battlefield conditions the stress rises, and under that stress decision-makers are far more prone to poor judgment, human error and impetuous choice in the absence of information. Indeed, under stress the most immediate tendency is to pay even more attention to the short run, fully discounting the future. Thus, if there is a second concept to be injected into policy-planning by each nation, apart from consideration and adoption of some general global principles, it could be this necessity to look ahead through a doctrine of anticipation—a collective look ahead, because what is at stake is our collective fate.

The primary focus of attention for these considerations has been the Law of the Sea. That conference has failed conspicuously to meet the need. It has catered to nationalistic ambitions and short-term benefits at a tactical rather than strategic level. Nowhere can there be found a technology assessment that inquires, "What may happen, if, or what may happen, unless, in one or two decades, with the Draft Treaty, without it, or with some alternatives?"

SUMMARY

Human activities everywhere on the planet are conspicuously affected by

**Margins for Survival - Overcoming Political Limits in Steering Technology,* by the author. Pergamon Press, Oxford, 1979.

the pervasive marine environment. Moreover, the role occupied by the oceans in the destiny of humankind and of individual nations is both energized and steered by politics. So far these politics have been based largely on historical tribalism, of a concept of national security, state by state. One of the fundamental realities, however, is that few activities of modern society are as clearly global in scale as those which involve the sea. The oceans are essential to the very existence of life on the planet. Geographically the oceans wash the shores of over one hundred separate nations. Oceans are a significant factor in their social and economic life, as well as their hope for progress.

At the same time, the oceans sustain a web of interdependencies among people that reflects two unspoken truths. First, all peoples enjoy a common legacy and are subject to a common challenge for survival. Second, as progeny of past generations who innocently benefited from the sea, all peoples today have a moral obligation of stewardship for the generations ahead. These truths impose on everyone an obligation to consider the global repercussions of unilateral actions in marine policy.

Every nation should assume, at the highest level, the responsibility to advocate global principles, as the transcendent protocol for international marine accord, and as backdrop to the generation of domestic marine policies by each in the family of nations. Secondly, the development of new policies should be rigorously conditioned by a formal consideration of the long-term as well as the short-term goals and effects. Finally, we should encourage members of the marine community to continue to address broader issues of marine *policy*, as well as more familiar issues of marine *science*. During the last fifteen years marine policy has itself emerged from the narrow parochialism of many pioneering oceanographers, who approached the study of oceans as though the planet were uninhabited. Now we must consider policy developed in a context featuring crisis management, complexity, interdependence, political expediency, information overload, institutional tribalism, a litigious society, alienation, the difficulty in consensus building, and the loss of confidence in government. All of which create a landscape of traps to rational decision-making.

If marine policy is to succeed, it is essential to recondition our entire policy-design process to face the future on a far more sophisticated basis than simply working within the narrow boundaries of marine policy. Most important is the need for a holistic, unparochial, future-oriented approach that accords with the dynamics of a modern technological society.

By so doing, those engaged in marine-policy design would have the satisfaction of two accomplishments. The first would be successful achievement of the social goals to which policies were directed. But the second—a far more trenchant and salient effect—would be to contribute consciously to new efforts for nations to use restraint in their unilateral initiatives, providing an opportunity for the oceans to serve as a rehearsal stage for world comity, meeting the challenge for survival in the 21st century.

CHAPTER 2

Regional Seas: An Emerging Marine Policy Approach

STJEPAN KECKES

Regional Seas Program
United Nations Environment Program, Geneva

INTRODUCTION

In accordance with resolution 2997 (XXVII) of the United Nations' General Assembly, the UN Environmental Program (UNEP) was established "as a focal point for environmental action and coordination within the United Nations system." The Governing Council of UNEP defined this environmental action to encompass a comprehensive, transectoral approach to environmental problems which should deal not only with the consequences but also with the causes of environmental degradation.

In the area of "Oceans," which the Governing Council identified as one of UNEP's priority areas, UNEP is attempting to fulfill its catalytic role and deal with the complexity of the problem in an integrated way, as demonstrated by UNEP's Regional Seas Program. Although the environmental problems of the oceans are of a global nature, it seemed realistic to adopt a regional approach, since in this way UNEP could focus on specific problems of the highest interest to the states of a given region and, therefore, could more readily respond to the needs of the governments and mobilize fully their own resources. By including activities of common concern to most coastal states, in due time this regional approach should yield a mechanism that will deal effectively with the environmental problems of the oceans as a whole.

PROGRAM FUNDAMENTALS

Cooperation with the governments of the regions. Since any specific regional program is aimed at benefitting the states of that region, UNEP tries to involve the governments in the program from the very beginning through their participation in the program's formulation and approval. The actual implementation of an adopted program is carried out through national institutions selected by their governments. UNEP's financial support always rests on the assumption that the governments of the region will themselves progressively cover the operating costs of the program as UNEP's initial catalytic role is fulfilled.

Coordination of the technical work provided through the United Nations system. Although the regional programs are implemented predominantly by the national institutions designated by the governments of the region with UNEP acting as an overall coordinator (in some cases UNEP's role is limited to the initial phase of the activities), a large number of the United Nations specialized organizations provide assistance to the national institutions, thus contributing to the program the support and experience of the whole United Nations system.

The substantive aspect of any regional program is outlined in an "action plan" which is formally adopted by the governments before the program enters an operational phase. All action plans are structured in a similar way, although the specific program for any region will be dependent upon the needs and priorities of that region. A typical action plan includes the following components:

Legal component. In most cases a legally binding regional convention, elaborated by specific technical protocols, provides the legal framework for cooperative action. The legal commitment of governments clearly expresses their political will to deal individually and jointly with their common environmental problems.

Assessment component. All programs include a large number of activities aimed at assessing and evaluating the causes, magnitude and consequences of the environmental problems. This assessment is not restricted solely to marine pollution but also covers the assessment of the coastal and marine activities and socio-economic factors that may influence, or may be influenced by, environmental degradation.

Management component. The assessment of the environmental situation is undertaken merely as a tool to assist national policy-makers to manage their resources in a more effective and sustainable manner. Therefore, each regional program includes a wide range of activities in the field of environmental management. Such activities may include co-operative regional projects on rational exploitation of marine living resources, utilization of renewable sources of energy, management of fresh-water resources, protection of soil from erosion and desertification, development of tourism without ecological harm, mitigation of environmental damage associated with human settlements, and the like.

Institutional component. The program is implemented primarily through designated national institutions. Assistance and training are provided where necessary to allow national institutions to participate fully in the program. Existing global or regional mechanisms are normally used for the effective coordination of the program. However, specific regional mechanisms may be created if governments feel it is necessary.

Financial component. As a program develops, the governments of the region assume a progressively increasing financial responsibility. Government financing may be channelled through a special regional trust fund or provided directly to the national institutions participating in the program.

At present there are eight regions where regional action plans are operative or are under development: The Mediterranean (adopted in 1975), the Red Sea (adopted in 1978), the West African Region (under development, expected to be adopted in 1980), the East Asian Seas (under development, expected to be adopted in 1980), the South-West Pacific (under development, expected to be adopted in 1981) and the Wider Caribbean Region (under development, expected to be adopted in 1980).

ACTIVITIES OF ACTION PLANS

It is extremely difficult to summarize the variety of activities which have been undertaken in the various regions, either as parts of the preparatory activities for the intergovernmental meetings called to give final approval for the action plans or as parts of the adopted action plans. Examples, selected very subjectively, highlighting our successes and failures, difficulties and pleasant surprises, could be presented as follows:

1. States, representing people with a different historical background, political system, level of economic development, religion, language, race and social goals, were willing to join their efforts to save the seas which they considered their shared heritage.

2. None of the meetings, either on the level of marine scientists, economic experts, planners, or plenipotentiaries, have ever raised politics above business to protect the environment. States at war were ready to sit together and forget, at least for a moment, their political differences.

3. Two regional conventions for the protection of the marine environment have been signed (Barcelona, 1966; Kuwait, 1978) and entered into force after they had been ratified with speed unprecedented in the history of international treaties.

4. Specific protocols, attached to these conventions, have been negotiated, signed and ratified, regulating dumping from ships and aircraft (Mediterranean and the Persian/Arabian Gulf).

5. Additional protocols are being negotiated, although the enforcement of some of them (control of pollutants from land-based sources in the Mediterranean) might cost the states concerned up to ten billion dollars.

6. A regional oil-combatting center has been established in Malta (1976) and the establishment of an emergency mutual aid center in Bahrain will soon be negotiated. Similar arrangements are envisaged for other regions too (South-East Asia, West Africa).

7. The marine environmental problems of many regions have been analysed through a series of international workshops attended by scientists from the region: Monaco (1974), Trinidad (1976), Penang (1976), Abidjan (1978), Santiago de Chile (1978).

8. A network of eighty-three national institutions was established in the Mediterranean (1975) to cooperate on seven pollution-monitoring and research projects using commonly-agreed methodology. Similar networks will soon start operating in other regions, all using methodology yielding results comparable on a global scale.

9. A large training program is under way to improve the quality of data generated by the institutions participating in the cooperative programs. A considerable number of sophisticated equipment (atomic absorption spectrophotometers, gas chromatographs, and the like) was donated to institutions in developing countries. A common maintenance service is organized for this equipment. Quality control of data is ensured by a mandatory intercalibration exercise.

10. The facilities of the Geneva-based United Nations International Computing Center were selected as the central data repository and processing facility. Data are collected, handled and disseminated according to existing, standard practices, making full use of the existing mechanisms for data exchange.

11. Regionally applicable environmental quality criteria are under development (Mediterranean bathing waters and seafood) and will be used for harmonizing national legislations.

12. The land-based sources of pollutants are systematically analysed and the amounts of pollutants reaching each region are estimated.

13. In a variety of fields related directly, or indirectly, to the protection of the marine environment (aquaculture, wetlands, marine parks, soil protection, soft energy, tourism, human settlements) intercountry programs have been initiated.

14. Attempts to get the governments' agreements for studies on the riverborne load of pollutants, on input of pollutants from the atmosphere, or for large scale modelling, have failed so far.

All these activities are based on the very close cooperation of fourteen specialized organizations of the United Nations system, coordinated centrally by UNEP.

CONCLUSION

UNEP's role in the development and financial support of the regional programs rests on the assumption that they will gradually become self-supporting and that the major responsibility (financial and substantive) will be transferred to the governments in the region. In the regions where the regional conventions, providing the legal framework for the regional action plans, have been ratified, this is actually happening: the weight of the governments in deciding on the program of work is increasing and the financial burden is being taken over by regional trust funds controlled by the governments contributing to them. UNEP's catalytic role in these regions will be mainly restricted to safeguarding the comparability of data, thus facilitating the assessment of the causes, sources, levels, effects and trends of marine pollution on a global scale.

TWO

The State of the Art

Timothy M. Hennessey
University of Rhode Island

CHAPTER 3

Providing Direction to the National Ocean-Policy Research Effort

ROBERT L. FRIEDHEIM

University of Southern California
United States

INTRODUCTION

From April 13 to April 16, 1978, twenty-six invited participants met at the Catalina Marine Science Center of the Institute for Marine and Coastal Studies, University of Southern California, at Big Fisherman's Cove, Catalina Island, California. The workshop was convened to examine the state of the art in national ocean-policy studies being done in academic institutions today, and to make recommendations to the Marine Science Affairs (MSA) Program of the International Decade of Ocean Exploration (IDOE) about inviting proposals on national ocean-policy problems. In particular, the workshop was to recommend attention: to problems which seem most important and amenable to research; to methods of analysis likely to lead to the best possible analytic results; and to the needs of "users" who would find national ocean policy research most beneficial.

The goals of the project were:

> First, to identify "the state of the art" in national ocean policy studies. This would include an examination of the literature with conclusions to be drawn as to its geographic scope, quantity and quality of coverage, processes examined, methodologies used, and trends over time.
>
> Second, to attempt to develop a consensus among producers and consumers of knowledge about national ocean policy as to the best directions for ocean-policy studies for several years into the future. That is, to attempt to obtain the best advice as to "payoffs".
>
> Third, to make recommendations to the MSA Program as to whether MSA should involve itself with national ocean-policy studies and which type

of studies should be supported. The criteria for those studies which might be recommended for support were those which promised improvements in the state of the art, filled in identified gaps in the literature, and helped decision-makers understand the social implications of new ocean knowledge.

Fourth, to attempt to develop a general framework for ocean-policy studies that would link these studies to larger societal concerns. Such a framework does not now exist.

Fifth, to strengthen the ties of the ocean-policy analytic "community" so that, through examination of a full range of work produced by ocean-policy analysts, better coordination and more rapid learning at reduced cost could occur. It was also hoped that "synergism" would develop; that is, that the cumulative product of the efforts of a number of analysts would add up to more than a mere sum of its parts.

The organizing paper and workshop discussion of the "state of the art" were attempts to realize the first goal above. Discussion at the workshop was designed to elicit from participants their recommendations (goal 2 above) for future national ocean-policy research, and hopefully lead to consensus concerning recommendations. The present chapter is intended to carry out the third goal, namely, to make recommendations concerning MSA's role in future ocean-policy research, and to assess which types of projects might have the maximum pay-off. Achievement of goal four was another motivating factor. Finally, participation and discussion at the workshop furthered the possible achievement of goal five—strengthening the ties, and therefore the learning, of the ocean-policy "community".

BACKGROUND AND MOTIVATION FOR THE WORKSHOP

The workshop and report on national ocean-policy studies sponsored by MSA came in response to a perceived need on its own behalf, and on that of the scientific community, the US government as a whole, and the organized collectivity of states—the United Nations.

The IDOE Program of the National Science Foundation—which sponsored the MSA effort—is "big science". It was begun in response to two needs. The first was a need for large-scale scientific efforts to understand the nature of the salt-water system, which covers almost three-quarters of the earth's surface. In addition nation-states, in political bodies such as the United Nations, demanded improved knowledge so that the world's people could make more effective use of the earth's resources. Ocean knowledge for its own sake and for applied uses was in short supply and unevenly distributed. Of necessity, the project's work would have to be large in scope, expensive, and trans-national. The pay-offs to mankind were thought to be immediate and evident. Much hope was expressed in the United Nations by US delegates that IDOE would show the world "what resources the seabed holds" as well as "how and at what cost these resources (can) be retrieved and used for human benefits."[1] Ten years of scientific effort has yielded a splendid harvest of basic scientific information concerning environmental forecasting, environmental quality, seabed assessment and living resources. But we still know little about the social implications of our new knowledge.[2] Part of the reason, it was thought, we knew so little about the possible consequences of new fundamental knowledge was that too little research money was spent on "basic research on the social and behavioral dimensions of man's relationship to the sea."[3] The MSA Program was established in response to the perceived need for more social-science research on man's relationship to the sea in general, and on the relationship of the new fundamental knowledge of the oceans being generated by the IDOE Program to implications for public policy. MSA was also authorized to consider projects that could improve the conduct and management of the IDOE Program.

In the first two years of the MSA program, proposals for social-science research that could meet MSA's mandated interests were found lacking in quality and quantity. Several reasons were alleged for the disappointing response to a call for proposals. First was the lack of an "identifiable" marine-affairs research community. Second was the unwillingness of many social scientists to transcend the usual boundaries of their disciplines and apply their skills to ocean problems. However, the third and fourth reasons for disappointment are probably more telling. The guidelines for MSA proposals were restrictive and required skills, methodological sophistication, and substantive interests rarely found in the right combination in existing social scientists. Linked to this is the lack of a "tradition of study that attempts to link new scientific or technical knowledge to policy implications."[4] This suggests there was little probability of achieving its goals unless MSA actively helped to create a community of scholars, widened its criteria for submission of proposals, and reoriented part of its program. Since IDOE's work is transnational, it does concern how nation-states are making and will make decisions relating to the allocation of ocean resources. Perhaps a useful place to begin to sponsor promising "social-impact" studies is to examine the context of these national ocean-policy decisions.

It is difficult to forecast precisely the "social-impacts" of marine policy, especially when extracting from a large body of data on the operation of the international political system and its national subsystems. Size, complexity, and its dynamic nature make it a formidable task under best circumstances. Under contemporary conditions it is even more difficult. The general structure of the system is changing rapidly. The most general manifestation of that change is what we all now recognize as a trend toward the "enclosure" of ocean space. Enclosure affects the system as a whole, as well as most of the major nation-state components of the system. In many major ocean-using states there is another important change which, many would argue, is best viewed as a manifestation of the same causes that move us toward ocean enclosure—a search for a "coherent" ocean policy. Both of these developments are now sufficiently evident, so there is a cadre of interested researchers. They are examining how the states of the world might respond to management of human activity on ocean space under a new regime, and evaluating the options for "coherent" ocean policy. A summing-up at this point could be useful both for the researchers and the potential research sponsor.

Governments with ocean interests have recently begun to accord priority to understanding what they should do and how they should proceed on questions of ocean policy. For example, US ocean policy was reviewed in 1968 by the Stratton Commission.[5] More recently there has been a spate of efforts to evaluate the initiatives undertaken in the late 1960s and set a new course, if necessary. The National Advisory Commission on Oceans and Atmosphere has made a series of recommendations.[6] A major study sponsored by the Department of Commerce, in which much of the US ocean effort is reviewed, has recently been published.[7] Finally, the President's Reorganization Project is believed ready to recommend a major organizational overhaul of U.S. ocean-policy machinery.[8]

MSA's purpose is consistent with these efforts, but much more modest. A number of scholars were assembled to ask if our ocean-policy knowledge was adequate to provide the necessary intellectual base for wise decisions by the next generation of governmental ocean-policy decision-makers. It made no sense to assemble a group of knowledge producers—mostly from universities—and knowledge consumers—mostly from governmental research-management offices—to recommend what current research may pay off immediately in government policy. But if the oceans are to have a future, we must know more about the evolving system which provides the context, about the behavior patterns manifested by various interest groups, political parties, governments, and international organizations, and about the policy options available. Basic ocean-related *social-science* research is as needed as basic "hard" science research if we are to understand—and perhaps master—the ocean world.

THE STATE OF THE ART IN ANALYZING NATIONAL OCEAN POLICY

What is "national ocean policy"? An acceptable, formal academic definition is "a set of goals, directives, and intentions formulated by authoritative persons and having some relationship to the marine environment."[9] It would include all activities relating to the substance of nation-states' uses of the oceans, how they make their decisions, and how they organize themselves to make their decisions. But many critics ask, is there anything in the nature of salt water which makes it sensible to attempt to treat policy relating to the ocean as a separate and distinct entity? This is often followed by the statement that there is no "coherent" policy for all land-related policy matters. While we would not claim that salt water per se makes ocean policy unique, we do believe there is a distinctive physical attribute to salt water that makes policy related to it have some important shared characteristics. The ocean—salt water—is a physical common. It is difficult to divide it up. It is difficult to prevent the currents from moving water where nature intended it to move, or fish to migrate where they will. Until recently we used all of the assets of the ocean, beyond the territorial sea, in common. Recently there has been a move toward enclosure—extending national jurisdiction outward from the coast, with varying degrees of authority, up to two hundred miles. While difficult to divide it up, it seems not impossible. But the common physical attribute of the ocean still creates some policy problems for humankind, even after the formal claims to jurisdiction are conceded. The oceans—like the air—are subject to multiple uses, some of them possibly adverse to the "owners's" wishes. All users are affected by this condition when they attempt to manage the uses of the ocean.

The universality of the two-hundred-mile enclosure movement argues that *de facto* national ocean policy already does exist. Since two-hundred-mile zones encompass most of the congested portions of the world's marine transportation routes, the most productive fisheries, all existing offshore oil and gas production facilities, and most marine recreation, there is little left in the oceans excepting polymetallic nodules and nuclear submarines for the next generation of policy-makers to concern themselves about. The myriad problems being tackled by governments have an underlying similar basis. Comparative analysis might provide useful insights.

A survey was conducted to assess the "state of the art" in studies published since 1967 that concerned themselves with the substance and process of national ocean policy. Eight "sectors" were chosen for special emphasis:

1. Organization and structure of National Ocean Policy
2. Ocean environment and Coastal Zone Management
3. Ocean Research and Engineering Development
4. Ocean National Defense and Policing
5. Fishing
6. Ocean Minerals
7. Ocean Energy
8. Ocean Transportation, Port and Harbor Management, and Shipbuilding

While the survey emphasized scholarly work in these fields a conscious attempt was made not to pull policy concerns completely from their contexts. It was recognized, for example, that many determinants of ocean minerals policy are non-oceanic considerations built into a national minerals policy.

Coders examined four thousand to six thousand items, of which 966 dealt with national ocean policy. In addition to the usual bibliographic information, we recorded: country, marine region, UN regional or caucusing group, which part of the decision "system" the work dealt with (e.g., input, process, output), process (e.g., legislative, executive), central, regional, or local level, methodology used, and sector (e.g., ocean minerals, fishing). We also attempted to discover which works included policy recommendations and which attempted to forecast a future state.

The findings are summarized below:

1. We know most about the *output* stage of ocean policy relating to all states surveyed. We know much less about *input* to ocean policy, and least about the *process* by which ocean policy is made.

2. Of the limited number of studies concerned with the decision process, most deal with *bureaucratic* decision. The literature on *legislative* and *high executive* decision is small. There is virtually no literature on *judicial* decision. However, the trend on all process-related studies is upward.

3. Some of the literature is "trendy." There was a sharp rise in ocean-policy output studies associated with the 1973-74 UN Law of the Sea Conference, then a decline, and then a more recent rise. However, there has been a steady rise in input studies, and possibly a sharp rise in process studies (a small sample size makes the observation less than definitive).

4. We know most about United States ocean policy. The next two most recently studied ocean policies are those of the USSR and Japan. Western European ocean policy is less known, and most Third-World countries' ocean policies have been little studied.

5. Sixty-six percent of all the literature concerns intra-political-group ocean problems. We know most about North America, less about Western Europe, and much less about Latin America, Africa and Oceania.

6. Thirty-one percent of the works surveyed dealt with specific marine regions, in the following rank order of availability of work: North Pacific, North Atlantic, Caribbean Sea, North Sea, Mediterranean Sea, South Pacific, Arctic, Central Pacific, Indian Ocean, and the Antarctic.

7. As to sectors, we know most about national ocean-policy organizational problems, fisheries, coastal-zone management, pollution, ocean minerals, ocean-related constabulary, and security problems. We know much less about policy problems of ocean science, engineering, transportation, and ocean energy.

8. We discovered that a good proportion of the literature is "policy oriented." Thirty-six percent of the studies made policy recommendations. In addition, twenty-eight percent of the authors attempted to forecast a future state, but about half of them did so with tools no more precise than legal, historical, or

first-person-experience modes of analysis.

DISCUSSIONS AT THE WORKSHOP

In some twelve hours of meetings over two days, discussion among the participants ranged widely over the organizing paper, the substance of national policies, the "system" of states, the most appropriate organizing notions, methodologies that might yield fruitful results, and recommendations to MSA for a useful program. No consensus emerged on any of these subjects, though there were some definite trends of support for a limited number of approaches to the problem of what should be studied, and what types of studies MSA and other funding agencies should support. Obviously, workshop members did not produce a strictly-ordered list of priority-action items for MSA's benefit, but all participants felt that the discussion could produce a useful orientation to those who manage ocean-related policy and social science research programs.

Future workshops may benefit from our experience in attempting to formulate recommendations. As we discovered, unless presented with a list of researchable topics and asked to rank order them, it is unlikely that attendees can help a funding agency choose worthy projects. Too many potential projects exist when a subject as broad as national ocean policy is under consideration. Too many varied professional, methodological, and ideological interests exist to make the selection process smooth and relatively painless. The most painful effort of all is any attempt to make recommendations concerning which methodology will have the greatest "pay-off" for any given preferred study objective. Discussion at the workshop was at times acrimonious, not only between practioners of different social sciences but among practioners of the same social science. It appears that many flowers are determined to bloom.

Surprisingly, there was little discussion at the workshop on political organizational problems. How the US government should be organized is currently part of an extensive, very visible debate in the "ocean community" in and about Washington. Organizational concerns subsume the belief that certain organizational structures are more appropriate than others for achieving certain ocean-policy objectives, are more "rational," or are likely to be more "efficient." We must leave it to the reader to decide that this negative "result" was a consequence of neglect, an artifact of the mostly academic invitees who chose to attend, or a valid indicator of the comparative importance of formal organizational considerations.

Despite the diversity of opinions expressed in two days, a limited number of foci, basic thrusts, organizational notions—or, in one or two cases, what some participants might characterize as paradigms in the

Kuhnian sense—did emerge.[10] While we believe the eight approaches represent the full range of intellectual foci that might be applied to national ocean policy, there was greater support (but no consensus) for one of them. The world-wide enclosure movement will have an impact on virtually all national ocean policies, according to a substantial number of workshop participants. Indeed, two participants subsequently wrote a guest editorial for *Science*, pointing out the importance of the phenomenon for science, resource use, and environmental mangement.[11] Other approaches were often defended by more than one participant, but no other shared this degree of support. The eight foci for national ocean-policy study follow:

1. *Enclosure.* Many felt enclosure can be looked upon as the independent variable in future studies of the political decisions relating to oceans. The trend toward ocean enclosure is clear. What is not clear are the dependent variables, among them: the detailed working out of trends; the national variations on details (different cultural values, economic systems, and stages of development); the "workability" and "efficiency" of virtually all nation-states to implement and enforce their enclosure decrees; and the short-, middle-, and long-run consequences to nation-states, other participants in the international system, and the international system itself.

2. *Allocation of Scarce Resources.* Several participants viewed ocean policy as one more arena in which the foremost criterion for analysis is using resources to meet efficiency standards. Scarcity demands as much. Some members felt we have too often espoused ocean goals without measuring benefits in relation to costs.

3. *The Ocean Policy Component of the Regulation of Post-Industrial Society.* The general social structure in the post-industrial age has put pressure upon scarce resources and fragile areas like oceans. Proponents of this view see the task of ocean policy as the development of sensible policies to avoid overexploitation. National ocean-policy studies should teach decision-makers to manage scarce resources in an era of limits. Coastal zones are where the current major problems occur, and studies ought to be emphasized there.

4. *Oceans as a Sector of National, Political, Economic, and Social Systems.* Ocean-policy concerns indicate that oceans are a subsector of national systems. That alone justifies their study. Some of the normal tools of political science can be employed if ocean policy is isolated out of the national system. Country or area studies can be performed. Comparative analysis of countries or regions can provide useful insights. Scholars in this tradition can contribute years of specialized training into different cultures, values, ideologies, and languages, and provide insights not usually obtained through other approaches.

5. *Economic Equity, or An Emphasis on the Distributional Aspects of Ocean Policy.* Several participants noted that the rules of the previous international ocean system—which appeared to be equitable—have had inequitable consequences. The major ocean-using states, they said, captured a disproportionate percentage of the oceans' economic value. National ocean-policy studies should be viewed as an opportunity to correct historic injustices.

Ocean-policy studies should be normatively based, not only because it will be just, but because those formerly-dominated,weak states are successfully forcing a change in their status. Resource-management problems are central to their concerns. Indeed, through the developing states' attempts to apply New International Economic Order criteria to ocean-mineral management in the UN Law of the Sea negotiations, we see what some would call the first post-war attempt to subject a resource problem to a common development perspective.[12] Ocean-policy studies of *developed* states should emphasize international equity; ocean policy studies for *developing* states should emphasize "catch-up" strategies.

6. *Building An Ocean Policy from Scratch, or the Dilemma of the Developing Countries.* A number of participants pointed out that regardless of how the developing fell behind the developed, the developing form a special case in national ocean-policy studies. Although many developing states have ocean interests and make decisions affecting their own citizens and those of other states, they rarely have an identifiable "ocean policy." In many cases, developing states must begin from scratch. Since their situation is substantially different, their ocean-policy system should perhaps be quite different, too. Whether this is so will be a fruitful area for research. However, since physical parameters constrain some elements of ocean policy for everyone, the developing may learn from the developed.

7. *Problem-Solving.* Despite limitations imposed by the academic schedule, some participants felt that university-based scholars should be asked to contribute to research. These participants were primarily concerned with "how-to" projects. For them, it was sensible to ask purveyors of fundamental knowledge to help find solutions to practical problems. Recommendations for typical research projects included: how to restore the US Merchant fleet to the status it enjoyed in the clipper-ship era; how best to manage the two-hundred-mile fishing-conservation zone; how the US should use its lead in ocean knowledge to gain an advantage in the future ocean exploitation. National decision-makers need this research to make use of the ocean opportunities unfolding.

8. *Inductive Approval.* Some participants felt something as large as national systems could be approached inductively—moving step by step, from the particular to the general—rather than dealing with the general

per se. Microanalytic studies at the local and regional levels should provide insights which could later be aggregated into a meaningful picture of the whole. Types of studies suggested included analyses of the policy impact of new scientific knowledge, and studies of ocean-using groups like fishermen.

CONCLUSIONS AND RECOMMENDATIONS

Although the group was large and disparate, a considerable degree of agreement (but not consensus) did form on the following questions: whether national ocean-policy studies should be pursued; whether MSA, IDOE and other funding agencies should have a role in promoting national ocean-policy studies; what the general purpose of national ocean-policy studies should be. We agreed on the eight foci listed above, on some suggestions for alternate funding strategies, on some considerations in evaluating enclosure proposals, and finally on a caveat concerning the emphasis on "national" in *national* ocean-policy studies.

National Ocean Policy As Focus

1. It was agreed that an important focus of ocean research should be the policies, decisions, and alternatives being pursued by contemporary nation-states regarding the use and management of ocean space. This would include *how* and *why* decisions are made, as well as the evaluations of costs, benefits, and impacts of ocean decisions, particularly those that affect and are affected by new knowledge of the natural world.

Marine Science Affairs Program

2. It was recommended that MSA fund studies in the area broadly identified as national ocean-policy studies. These are valuable in and of themselves, but also provide the context for studies of the direct impact of new knowledge. No other funding agency within the US Government emphasizes such studies.

3. Because there is no prospect that MSA-program funding will grow significantly, it was recommended that other funding agencies share that responsibility. Subjects to be examined under the national ocean-policy label touch virtually all government agencies. Previous studies have demonstrated that money spent to understand the social, political, and economic aspects of ocean use is well out of balance with that spent to understand the physical aspects.

Purpose of National Ocean-Policy Studies

4. It was agreed by virtually all participants that the purpose of a program to sponsor national ocean-policy studies was to improve the ana-

lytic base for advocacy. The intent of the program should be to prove the best policy analysis.

Approaches

5. It was agreed that all of the eight foci described above could provide useful analytic results. The variety reflects both different perceptions of *need* as well as different *methods.*

Alternate Funding Strategies

Although participants reduced to eight the foci appropriate for analyses, they could not achieve consensus on the primacy of any one. We can report only four suggestions for priorities in funding strategy, in descending order of support:

6. Place the emphasis upon causes, process, impacts, and consequences of enclosure efforts by the present generation of nation-state decision-makers. This funding strategy was supported by a majority of participants, who claimed that ocean use for generations to come is being heavily influenced—if not decided—by placing under national jurisdiction areas up to two hundred miles from coasts.

7. The second-most popular suggestion was to concentrate studies into analyses—comparative in method—of the major ocean-using states: the United States, the USSR, Japan, Canada, the United Kingdom, France, Norway, and Brazil. These states are actively exploiting and degrading the oceans; the policy patterns they establish, the institutions they use, and the consequences of their practices are likely to affect many other states—both developed and developing—as they struggle to solve their own ocean problems in the following generation.

8. One cadre of participants thought the ocean problems of developing countries required some studies oriented specifically toward them. There was little disagreement on the distinctiveness of their problems, but less agreement on whether MSA should place special emphasis on such study.

9. Another stratey might be to fill the "holes" identified in the survey of the state of the art. With the exception of ocean transportation materials—where there may be a large, high-quality, specialized literature that was not identified in the survey—there was general agreement that the survey successfully identified work needed on the input, process, and output of national ocean-decision systems, on which states were not adequately covered, and on which sectors (e.g., fishing, coastal zone management) were neglected.

Enclosure

10. Since so much time was devoted to discussion related to the causes, effects, and consequences of enclosure, more useful information emerged on

this subject than others. Listed below are some considerations for judging a proposed study.

A. The proposed research is rigorous and thorough.

B. The proposal deals with a state that is actively pursuing enclosure rather than merely issuing enclosure decrees. The "indicators" of active pursuit include implementation costs, enforcement problems, a visible "politics" of enclosure, and allegations of important consequences to new and traditional users, domestic or foreign.

C. The project proposes to examine multiple-use problems in enclosure.

D. The project is designed to try to measure the impact(s) of enclosure, preferably with sophisticated tools or by methods which allow comparison of results.

E. The project helps clarify objectives that might be realized in enclosure.

F. The project deals with some aspect of enclosure that proposes to promote or retard the conduct of ocean science, or to help or hinder international scientific cooperation.

G. Since the purpose of the research is to improve the analytic base for policy advocacy, the project should concern state, marine region, or policy problems of interest to policy-makers.

Caveat

11. Although there has been fierce debate in the social sciences as to who are "actors" in the contemporary world political system, and whether their relative importance is shifting, most analysts would agree that the nation-state remains a primary actor. It makes sense to promote research that concerns *national* ocean policy. But nation-states are not the only actors on the world scene, and total involvement with the behavior of nations may mask the important role played by transnational public organizations, multinational profit-seeking organizations, transnational private organizations, and even individuals. Studies of national actors should be designed to take account of non-state influences as well as the international system.

NOTES

[1]Statement by Ambassador James Russell Wiggins, *Press Release USUN-192,* November 6, 1968.

[2]Lauriston King "The International Decade of Ocean Exploration Program in Marine Science Affairs," *Marine Technology Journal* 11:5-6 (1978), p. 10.

[3]Ad Hoc Sub-committee for the Interagency Committee on Marine Science and

Engineering. *Federal Agency Support for Marine-Related Social Science Research,* December 1976, p. 11.

[4]Lauriston King "The International Decade of Ocean Exploration Program," p. 14.

[5]Commission on Marine Sciences, Engineering, and Resources, *Our Nation and the Sea: A Plan for National Action,* Washington, D.C.: Government Printing Office, 1969.

[6]National Advisory Committee on Oceans and Atmosphere, Federal Reorganization, *Summary Report,* September 17-19, 1978.

[7]U.S. Department of Commerce *U.S. Ocean Policy In the 1970s: Status and Issues,* October, 1978.

[8]*Federal Register* 42:243 (December 19, 1977), pp. 63665-63669. *Ocean Science News* November 6, 1978, December 11, 1978, and January 1, 1979.

[9]John K. Gamble Jr. *Marine Policy* Lexingston: Heath, 1977, p. 8.

[10]Thomas Kuhn, *The Structure of Scientific Revolutions,* 2nd Ed. Chicago: University of Chicago Press, 1970.

[11]David A. Ross and Edward Miles, "The Importance of Marine Affairs," *Science* 201:4353 (July 28, 1978).

[12]For a discussion of this problem see: Robert L. Friedheim and William J. Durch, "The International Seabed Resources Agency Negotiations and the New International Economic Order," *International Organization* 31:2 (Spring 1977), pp. 343-384.

Commentaries

DOUGLAS JOHNSTON

Dalhousie University
Canada

In a sense, what we ought to be discussing is how the human intellect conceives of the ocean in the late twentieth century. My son in junior high school is at the stage of trying to distinguish the human being from the rest of the animals. According to his textbook, we are supposed to have a unique combination of charactertistics. No one of these is unique to man, but the combination is unique: rationality, imagination, the power to reason and to abstract, to plan, and so on. Let's think of ourselves for a few minutes as a unique species. How do we use the human intellect today, in what is acknowledged to be a hybrid area of human thought and knowledge, an area of comparative cross-disciplinary inquiry which brings together all sorts of diverse people trained in different types of prejudice and sophistication?

What are we to make of this new hybird? We don't have a name for it yet, for one thing. Some still call it "the law of the sea," but that doesn't sound quite right, now that ocean management studies have been added, as well as the terrestrial components of coastal zone management. Some people call it "comparative marine-policy studies." Others call it "ocean policy," and no doubt further alternatives occur to us. Whatever we decide to call it, it is different from anything that came before. Yet it came from somewhere, and perhaps we can look at its roots.

In my view at least, the human intellect was first applied to the sea – in the age of recorded history – for purposes of legend, mythology and literature, when the motivation was wonder, awe and reverence. With our focus on ocean policy in the twentieth century, we may have lost something of our human talent for awe and reverence. Most of us would like to think of ourselves as hard-headed, pragmatic, skeptical, and even cynical about the sea. Maybe the scientists among us are still capable of surprise and wonder in looking at the natural system of the ocean, but the social scientist – I would suggest (being one myself) – does not have this talent. There is a loss in this.

If we come forward beyond the age of drama at sea, we come to an extraordinary period of human history – the Roman Empire. We still think like Romans about the sea, which is curious in two ways. Firstly, the Romans were not really a seafaring people. The question in the late twentieth century is, what will be the impact of the fact that the majority of nations in the world today are not really seafaring peoples? Will their collective influence in providing a legal framework be as profound in the 21st century as that of non-seafaring Rome in earlier times? And will it be left to seafaring peoples to provide the legal framework with technological content? As you know, the international trade system built by the Romans in the Mediterranean depended almost exclusively on non-Romans. The Romans almost killed Christianity, among other things, and virtually brought on the Dark Ages by their lack of maritime skills. After his conversion in Rome, Augustine, the greatest intellect of his day, returned to North Africa. Soon, scholars set out to study at his feet, and the only way of getting to North Africa was by sea. Modern research suggests that one-third to one-half of them died at sea—on the way out—and one-third to one-half of those who arrived did not get home safely. Shipwreck and disease contributed to these losses but the cause above all others was piracy, which, despite its formidable military might and technological sophistication, the Roman Empire was unable to control.

The second curious feature of the Roman approach to the law of the sea was its emphasis on *status*. The Romans derived everything from the concept of status, and that essentially is how everybody—lawyer and non-lawyer alike—continues to think about the sea today. What I would call the *spatial* concepts underlying the classical law of the sea are derived essentially from Roman concepts of status. It is only in very recent years, if indeed at all, that we have begun to try to think legally about the sea in non-status, non-spatial terms.

A famous jurist of the nineteenth century, Sir Henry Maine, theorized that legal development proceeds from status to contract. Now that we are in the age of joint ventures, treaties, and other kinds of contractual arrangements, perhaps we are about to proceed to the contractual stage in the development of the law of the sea. Are we capable of making such a leap in our thinking about the sea, enabling transactional, organizational and managerial concepts to prevail over status concepts?

In the Dark Ages, after the Roman civilization expired, we went through a period of an almost totally non-intellectual approach to the sea, when the human intellect was simply not applied in any organized manner to the ocean. The sea was perceived essentially as a spatial vacuum, a convenience for transit and trade. For over a thousand years, ship technology scarcely advanced at all. Indeed, it is only in very modern times that we have made impressive advances in the ocean, with the arrival of fixed technology and dramatic improvements in mobile (vessel) technology. This was a long period of human stupidity, if you like, and for reasons difficult to understand.

Then came modern warfare, and intelligence asserted itself again for strategic reasons. Sea power became the focus. Overlooking for the moment some moral questions, we might be grateful for naval technology. It is intellectually interesting, if morally discouraging, that it was the military planners of the late eighteenth and nineteenth century who began thinking intelligently, if not nobly, about the sea. Most modern thinking about efficiency in ocean management—both in the form of specialized use and in that of versatile use—can be traced to the military planner.

We now reach the age of science. Ocean science, of course, is very young, and this complicates the way we think about the ocean. Science is rigorous, slow, and expensive. It sets up very rigorous evidentiary standards, more demanding than lawyers have ever tried to establish for determining guilt or liability in court. Vessel-based research in particular is almost sinfully expensive, and it takes too long for management decisions that have to be made. In the normal rush of events, we simply cannot wait for conclusive evidence. So un-science intrudes: impressionism, suggestivism, bigotry, tribalism. All these things that scientists abhor are with us in ocean management—in part, it is said, because scientists set too-high standards for normal human beings. But what scientists did was establish the notion that, with the best will and effort in the world, we might strive to make rational use of our ocean.

Science has also created concepts of specialized use that are now under strain, for the reasons Robert Friedheim has suggested. We cannot afford to remain within the existing compartments. These boxes may be rationally defensible from a purely scientific point of view, but we cannot afford them when policy decisions have to be made. Science has also contributed to the rise of jargon. Interestingly, however, now that we are forced to communicate across disciplinary lines because of our common interest in policy issues, jargon goes out the window. Most literature today on marine policy matters is written for conferences, workshops, and colloquia, so the need to speak to one another determines that we write in a manner comprehensible to all. Some precision is lost.

On to the twentieth century. Because of the way the Romans thought about the sea—as a great space in which everybody can roam around freely with a minimum of constraints—"internationalists" were the first to focus on ocean-use problems. Later, international lawyers were joined by specialists in international organization, by political scientists who specialize in international politics, and by diplomatic historians. Now we have a new mix of people who presumably will call themselves "comparatists."

In economics we may expect a "comparative" trend; specialists are tempted to compare different kinds of resources. Most of today's senior fishery economists came from forestry, which has led to analogies between marine and non-marine renewable living resources. The arrival of fixed technology, and the evidence of environmental harm caused by improper use of ocean resources, bring in concepts of social cost and raise the need to compare land-use and ocean-use analyses. The current emphasis on energy studies seems likely to introduce another level of economic strategy into "ocean economics." This new sub-set of the discipline seems bound to evolve into much more of a comparative study than seemed likely in the 1950s, when it seemed to consist only of fishery economics, on one hand, and shipping economics on the other.

The contemporary decision to divide up much of the ocean into national compartments is likely to have a profound effect on the way we look at the ocean in the 21st century. For one thing, the new emphasis on *national* policy means we look at *regulation* and at least the prospect of *management*. The shipping industry was the last of the great industries to be brought under some degree of regulation. The fishing industry until modern times was largely unregulated, and the fisherman indeed enjoyed being a fisherman partly because his life was thought to be too strictly regulated on land. He enjoyed being "free" at sea—a luxury that is now denied him. With the national policy orientation we now have, we are obligated to focus on the sociological side of regulation and management. This is already attracting the attention of the sociologist and the anthropologist, for the first time in marine affairs. Many of these specialists are emotionally set towards defending established life styles against what are perceived to be instrusions of the state and state organs in the coastal community.

Perhaps we are becoming less noble, less generous, less imaginative probably, in the way we look at the sea. I would like to think this is just a temporary deviation. We pride ourselves on being hard-headed, pragmatic, skeptical, and cynical, and are sure this style is going to prevail. This is a proper way of looking at things. Anything else is unrealistic. Yet, as we are driven to be comparative, we will be studying the force of imitation in human affairs. Man, as a species, is a pretty imitative animal. As we imitate one another we are probably engaging—despite ourselves—in a new kind of co-operation.

LAURISTON R. KING

Texas A & M University
United States

The rationale for pursuing comparative marine policy embraces problems of broad scope as well as some that are very parochial and self-interested.

First—in keeping with a notion of the doctrine of anticipation—we have a situation that is not critical, in terms of a need to find a solution to some threatening social or environmental situation. Marine affairs, as a field of study, does not yet bear this burden. It would be very difficult for a government official, I think, to argue that failure to mount a vast and expensive effort to better understand the global marine-policy process would result in dire consequences. That the oceans are not pressing upon us in the way that a number of other environmental areas are is important; it provides the luxury to think and to explore, and to conduct some research we only dimly perceive as important.

The timing is right for the work described in Robert Friedheim's program review. From a national perspective, a first benefit is strategic intelligence, in a politically neutral sense: understanding ocean policies as they begin to evolve in other nations. It is incumbent on the United States as a major maritime power to understand what solutions other countries are turning to, what problems they are trying to solve, how they are related to problems in the United States, and who is responsible for resolving those problems.

Secondly, we can refer to "enlightenment:" the simple satisfaction and inspiration that comes from understanding how others deal with similar problems, how others define problems, and how others use their institutions to set about resolutions. Such understanding greatly enriches our own perceptions and capabilities.

Thirdly, taking a comparative approach to the study of national ocean policy may develop a cadre of individuals who understand this area. I'm increasingly impressed by the way ideas and studies are transferred. Despite the amount of argument devoted to "what are you going to do with the study once you get it," knowledge generated in these studies moves mostly by individuals who have either performed the study or have been closely associated with it, and who then move into positions of public responsibility.

I think that's the main vehicle for moving new knowledge in a way that influences public policy. If we can develop and encourage this movement in and out of government—primarily through universities and imaginative approaches to internships—we can increase the flow of ideas. In order to do that we must have the basic intellectual foundations, but here we are seriously deficient.

Another dimension to this is that it provides certain economies for public agencies, reducing training time by getting people who already understand something about the international context. It's also a way to overcome certain regional parochialisms.

Comparative marine-policy studies can also provide a basis for the short-term studies agencies often need. Many ocean-policy studies have been contracted out on a very short-term basis, and done under great pressure without sufficient basic knowledge. This is particularly true for contracting agencies and other such agencies in the government. If the Library of Congress is directed to do a study, they must draw largely on existing knowledge. Secondary users pressed for applied studies have very limited resources available to them.

Finally, a critical strain has been markedly absent from much of the literature. Academics have a real responsibility here.

HARVEY SILVERSTEIN

University of South Carolina
United States
(presently Dalhousie University, Canada)

Implicit throughout Robert Freidheim's study was the suggestion that we need a common understanding of what we are doing and how we are going to do it, and what methods we are going to employ. I am going to suggest just the opposite. We should not come to common agreement and set limits on our profession at this point. The developments in marine policy are so dynamic, rapid, and changing that we might lose their vitality. I would like to discuss four points which bear upon this major argument.

Firstly, we need to communicate more with other disciplines, particularly with technologists and marine engineers. Technologists participate in our functions, but we have gone almost nowhere in participating in theirs. Yet when it comes to the issues—whether construction standards for tankers, or potential for ocean energy production—a great deal of the action is in engineering. We also need to solicit more anthropologists, psychologists, and sociologists to participate in the activities of the marine policy community.

We need to know more about how other countries perceive themselves, how they make their own decisions, what their objectives are, how they formulate and carry out their own foreign policy. We can make far better use of historians, linguists, and foreign area experts, who may know nothing about ocean policy but who are expert in foreign decision-making. Cultural differences of many types make for dramatic variation in basic values. In some societies religious obedience is far more important that logic or political expediency. It is important to recognize these differences in national perceptions and behavior. They yield marine policies very different from our own.

To a large extent policy formulation is not scientific nor objective but idiosyncratic. We have to recognize the role of individuals, and biographical or psychological studies could help show us how things actually happen in marine policies. Marine history is replete with examples of how one person turned a country around, initiated new technology, or moved off in a new direction. From Charles Darwin to Hyman Rickover, from Jacques Cousteau to Arvid Pardo, actions of particular individuals have brought about formidable changes. Social scientists have become too concerned with computerization or the "scientific integrity;" we have lost track of the idiosyncratic variable in ocean policy. Another point again brings me to some diagreement with Robert Friedheim. The emphasis on enclosure is short-sighted. History provides us with frequent examples of what can be called enclosure: under the Spanish Armada, under Britannia, and in the complex fishing codes of many countries.

The fundamental stimulus to ocean change is the role, implications and applications of new technologies. The Navy is using neutrino systems which go right through the body of the earth to detect submarines. The U.S. government is funding $60 million dollars to generate ocean energy off the coast of Hawaii. The Japanese have launched the "Kaimai," a new vessel which produces two thousand kilowatts from ocean waves. Technology is having dramatic effects on every policy consideration. We need to be aware of what is going on before we get hit in the face with it.

ARILD UNDERDAL

Nansen Foundation
Norway

From Robert Friedheim's point of view we might conclude much work is being done on ocean policies. I suggest the state of the art is actually much worse than a number of studies would indicate.

The study of marine policies is essentially a non-cumulative field. Except for economics, it is weak in theory, particularly in political science. One consequence of this is that we have a lot of descriptive studies unrelated to theoretical concepts, or where theoretical concepts are so poor it is hard to draw any conclusions. One such "concept" is "integrating" marine policy. What do we mean by "integrating" marine policy? How do we conclude that one system or one set of policies is more "integrated" than another? We must relate our descriptions to theoretical constants. The main problem is lack of theory, not necessarily the lack of comparative studies. If we had a common conceptual program or common theories, we might draw general conclusions from a lot of separate case studies that are not themselves comparative.

We must distinguish between studies that are themselves comparative and case studies which could be used for comparative purposes. Now case studies cannot be used for comparative purposes if methodology is weak. We in Europe are somewhat skeptical about an American political science professor zooming in for a fortnight, telling us how our political systems work. If we strengthen our conceptual background we can make better use of native scholars, and won't have to rely on this kind of superficial comparison.

This brings me to my final point. Improving communication has a certain advantage. On the other hand, what we are saying is not very interesting. In literature there are frequent references to "equity," which most people agree we should work towards. What we mean by equity seems a very crucial question. We could all probably agree to study how different institutions produce or do not produce equity, but before we invest a lot of energy into such a study, we should invest a small amount of effort and time to make clear to ourselves what precisely we mean. Before we request millions of dollars to study to what extent marine policies are "integral", we should perhaps spend $10,000 to make ourselves a bit more aware of what we mean by those principles, those concepts.

THREE

Coastal Management

Francis X. Cameron
University of Rhode Island

CHAPTER 4

The Coastal Policy Issue in Sweden

ERIK CASTEN CARLBERG

National Board of Physical Planning and Building
Sweden

My topics in this chapter are the overall conditions and problems concerning land-use planning in Swedish coastal and archipelago areas, the utilization of water areas outside Swedish coasts, and the utilization of physical resources in the sea and beneath the seabed.

MANAGEMENT OF LAND AND WATER RESOURCES

In Sweden, the comprehensive planning of *land* resources and *land* use has been incorporated into relatively stable political, administrative and legislative forms. Those concerning the use of *water* areas and the *seabed* still remain undeveloped and diffuse. National physical planning work has enabled the formulation of long-term, concrete guidelines for *land* use in our coastal areas (Figure 1).

During the last decade the development of joint collaboration between the Government and Parliament, the central government and local authorities, and between the general public and interest organizations resulted in considerable unanimity about the main forms land use in coastal areas should take. The problems primarily considered deal with the location of environmentally detrimental industries and power plants along the coast, as well as the problems connected with the growth of vacation-cottage development and recreational facilities and their relationship to fishing, nature conservation, the conservation of historical and cultural monuments, and the interests of permanent residents.

Land-use problems concerning coastal areas were dealt with by the Government in a special decision during the spring of 1979. This decision, discussed in Parliament later in the autumn, provides a long-term foundation upon which the conservation and management of nationally

Figure 4.1: The Geographical Guidelines

important land resources can be based.

Physical land-use planning has a long tradition in Sweden. Approximately two-hundred seventy local authorities are primarily responsible for land-use decisions. According to the Building Act, local authorities can determine the main outlines of land use in the master plan, as well as excluding particular areas from development. Where it is necessary to encourage land use of particular importance, the Government can draw up a master plan to meet national interests for a particular area.

The location of industrial or other activities important for conservation of Sweden's total land and water resources must also be referred to the Government. Regulations incorporated in the Building Act are fundamental to the division of responsibility between central and local government authorities, as well as different types of private interests.

COORDINATION OF MARINE RESOURCES

For the planning of the use of *water* areas and *marine* resources, there is a lack of traditions or clear-cut administrative and legal forms. Of course legislation and other ordinances valid on Swedish territory can be applied, where relevant, to coastal territorial areas. This is the case, for example, with the Building Act, the Water Act, the Nature Conservation Act and the Environment Protection Act. But a fundamental question regarding the geographical administrative sub-divisions thus stretches, in principle, as far as the limits of Swedish territorial waters. But boundaries between these different local-government areas have not been determined. (The Swedish Marine Territory is illustrated in figure 2.)

During recent years interest has been increasingly directed to the marine environment and its natural resources. Conflicting interests between different forms of utilization have become apparent and have increased the need for knowledge about the marine environment, as well as the planning of the use of sea areas. Several bills have come before the Swedish Parliament concerning the need for protection and conservation of the marine environment.

A special working group on marine environment problems has therefore been attached to those involved in national physical-planning work. The group is to collate and discuss existing information of use in the planning of the protection, conservation and utilization of marine areas. One result of the group's work thus far is illustrated in figure 3, showing those rather limited marine areas around the Swedish coast where obvious conflicts occur between different interests.

Last year the Government set up a delegation at government level for the coordination of problems concerning marine resources. The following quotation summarizes the background to this decision:

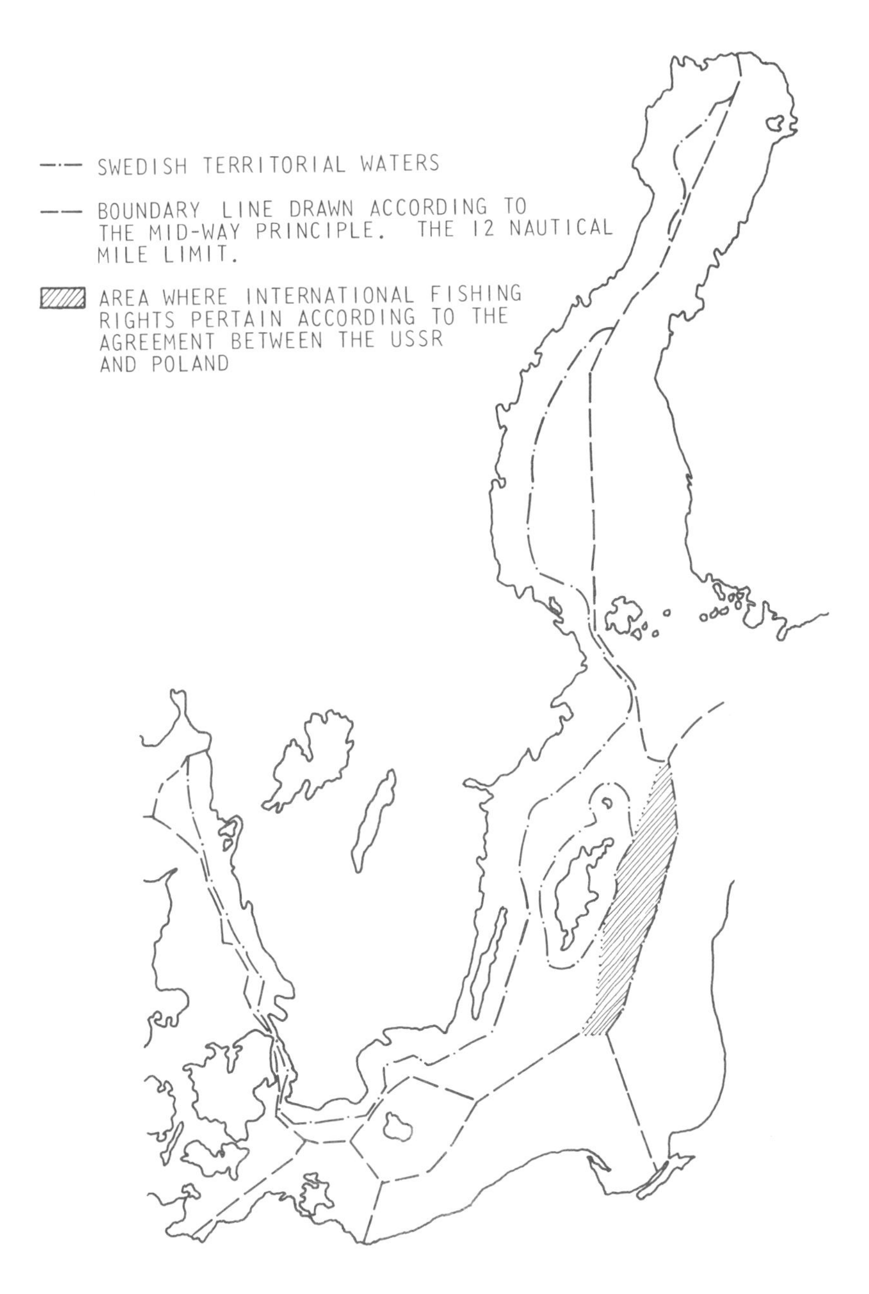

Figure 4.2: The Swedish Marine Territory

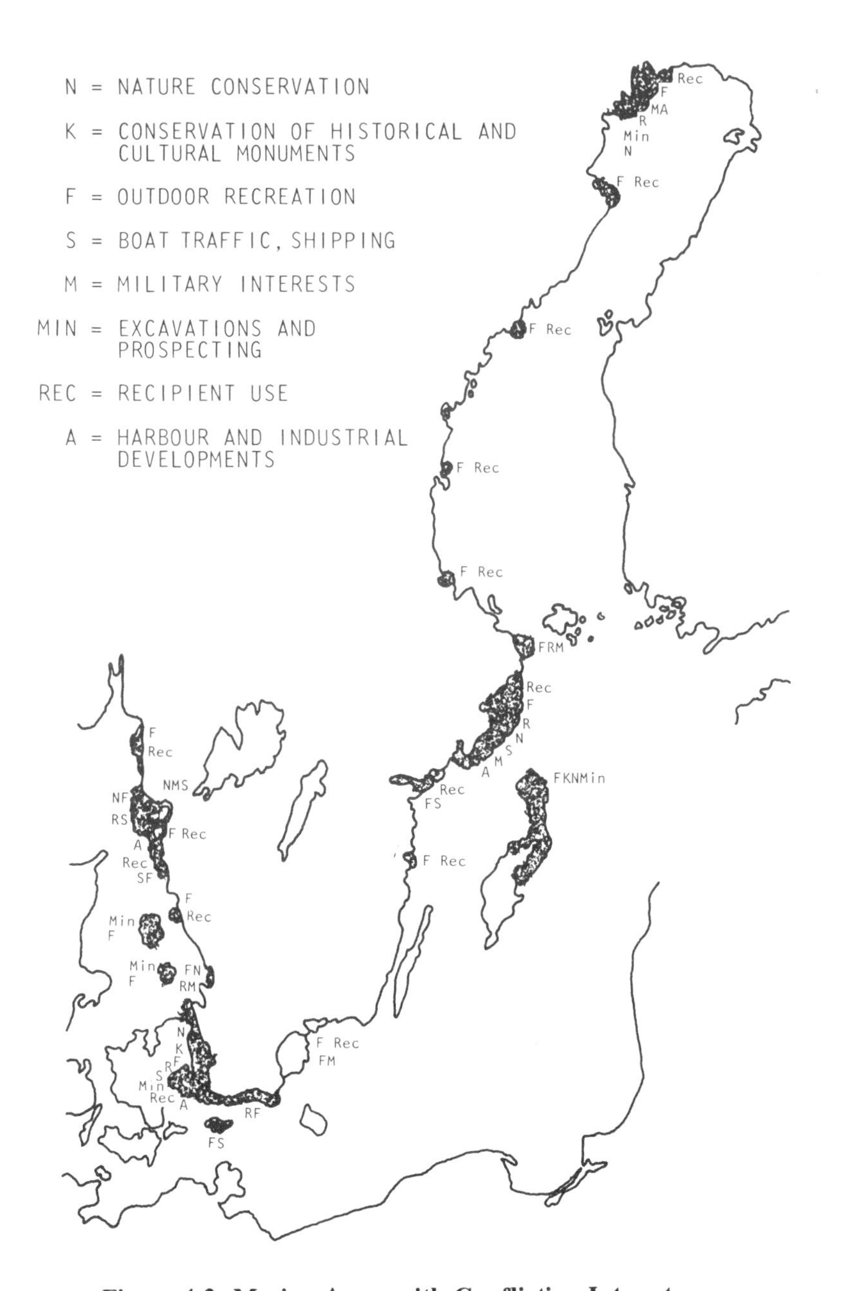

Figure 4.3: Marine Areas with Conflicting Interests

> Out of fourteen Ministries, ten—to greater or lesser extent—have deeply involved marine resource interests. This naturally leads to considerable coordination problems. The Ministries represent and plan for their respective sectors and approve grants for subordinate bodies in accordance with their sector plans. In several fields there is a lack of administrative prerequisites for the coordination of joint planning activities in the long-term. Similarly there is also a lack of overall planning that is aimed at the conservation of marine resources and with its coordination with physical planning on land.

The Marine Resource Delegation had its first meeting in June 1979, to discuss the direction of its work. The Delegation's secretariat emphasized the importance of forward-looking activities and stressed the fact that Swedish marine resources not only include physical resources but also the availability of the sea: for transport purposes; for the extraction of energy; as a purification unit for sewage; and as a climatic regulator. In addition Sweden has an important resource in the sea for technology and organization—fishing, shipping, navigation, research and development, international contacts between authorities, and commerce.

With regard to the activities of public and official bodies, efforts will be required concerning environmental legislation, planning, and participation in international joint projects. With regard to commercial activities, it is assumed increased efforts will be made to develop new techniques and methods for activities in Swedish and international waters. Industrial activities in the sea are expected to increase extremely rapidly. Stringent requirements will be placed on the public sector in terms of the supervision, control and steering of activities so that other interests are not constrained.

The Marine Resource Delegation's secretariat has proposed that permanent working groups be set up to deal with the following overall themes:

1. International and national legislation, and what demands these might place on Sweden;
2. Marine ecology, and the effects of toxic substances and other discharges on the ecosystem;
3. The supervision system for marine areas;
4. Fish and algae farming, as well as other industrial activities in Swedish waters;
5. Swedish activities in international waters; and
6. Research and development, its organization, and the way programs are revised.

Joint consultation opportunities will probably be initiated between the Delegation and representatives of those working on national physical planning. A study will likely be made for the coordination of systems and methods for planning and decision-making.

National physical-planning work has provided a long-term background for decisions concerning short-term aspects of development, and particularly problems connected with industrial growth in coastal areas. For further out from the coastline, we need to revise forms for decisions and applications for the following types of development: discharges from ships; the laying-up of ships; bunkering; unloading; storage of materials dangerous to the environment; harbor and marina developments; regulation of boat traffic; activities within harbors; intake and discharge of cooling water; use of chemicals; underwater blasting; pipelines and other conduits in the water and on the seabed; excavations and prospecting; and dredging, tipping and dumping.

It is still unclear how experience gained from national physical-planning work can be used to plan for the use of marine resources. However, many things point to the county administrations, in consultation with local authorities, to the central-government bodies involved, and to private interests, all being given the task of surveying long-term problems in their respective marine areas.

MARINE NATIONAL PARKS

The National Environment Protection Board published a study concerning marine national parks, i.e., areas that should be protected because of their value for scientific and biological study. The Board suggests that thirty-eight marine national parks be established, representing six different marine environments: archipelago; open, shallow coastline; river/delta; bay/creek/bight; exposed island; and deep-water areas at sea.

The Board has had as its primary starting point *ecological* criteria, and has assessed marine areas according to their representative status, adaptability, exclusiveness, untouched state, value as genetic banks, and importance for the survival of species. As its second point it has had *scientific*, *production*, and *biological* criteria. These have been used for geo-scientific, hydrographic and marine-biological research; for educational interest on land; and for fishing.

In order to carry out this proposal, it is necessary that the Government and the Swedish Parliament take overall decisions as part of national physical-planning work. The core of these areas could be protected by the environment-protection legislation; buffer zones could be protected by planning decisions taken in the same way as for planning on land. The Board has suggested responsibility for management be delegaged to working groups in each County Administration.

In the study carried out by the Board, Swedish coastal waters were divided into six sections, according to—among other things—salinity, hydrography and coastal topography. The ecological conditions vary widely, as can be seen from figure 4. In the Gulf of Bothnia, only three

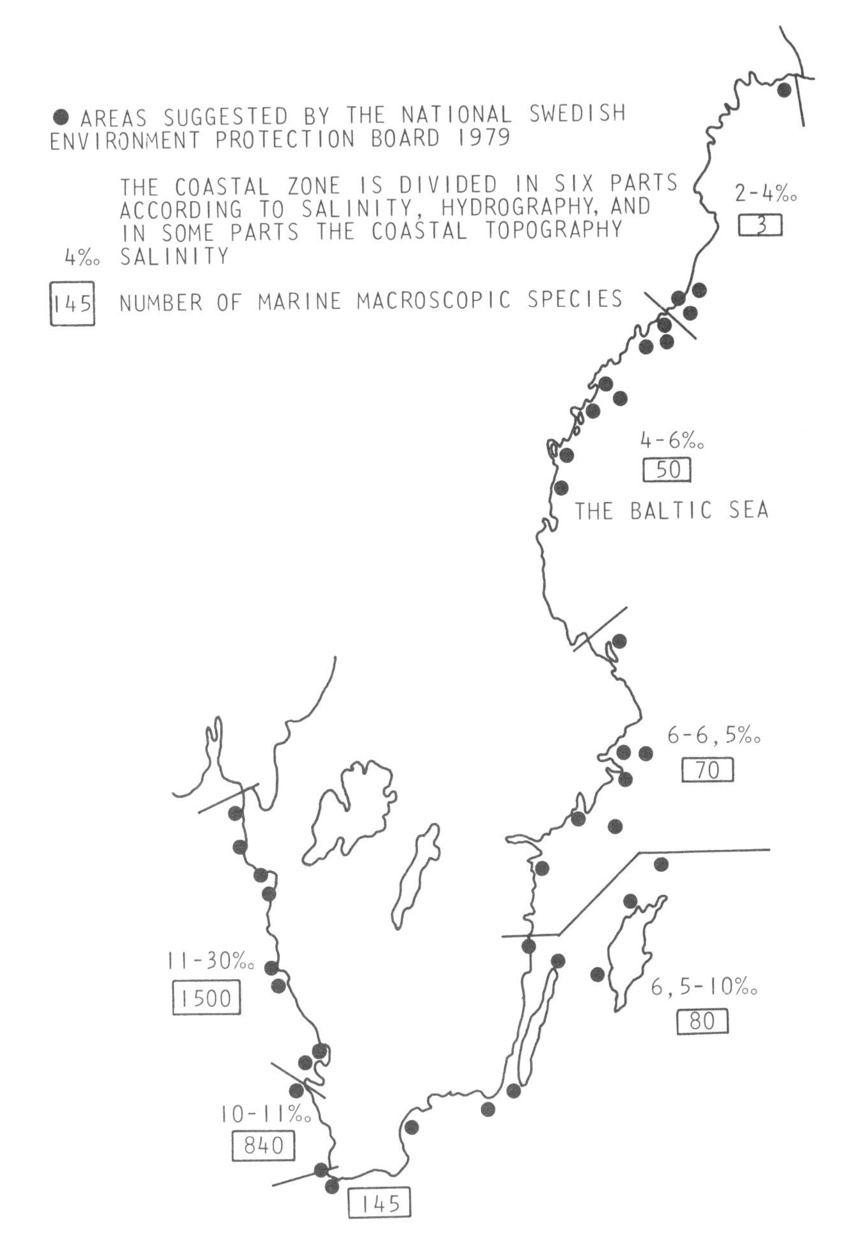

Figure 4.4: Marine Areas Suggested for Protection

macroscopic marine fauna are to be found, while in the Skagerak there are fifteen hundred. The study illustrates in a very direct way the need for revision of planning and decision-making systems for marine areas.

Principal guidelines need to be established at Parliament and Government levels. It is, however, likely that we have many years' work ahead of us before Parliament's guidelines for the conservation of *land* and *water* resources will actually include our *coastal* and *marine* areas.

CHAPTER 5

Coastal Zone Management in the United States: A Characterization

MARC HERSHMAN

University of Washington
United States

ACTORS AND FACTORS IN CZM

Many governmental agencies in the United States are concerned with Coastal Zone Management (CZM) and we often describe it in terms of a particular one. In order to do effective research, we must characterize CZM with some precision. Before listing characteristics, though, we should take a look at the actors.

Figure 1 provides a general schematic of the primary agencies concerned with CZM. It follows the line of principal political organizations. At the federal level we have three sets of actors: the Corps of Engineers review process; other federal agencies; and the CZM Act. At the state level we have a fourth actor, and at the local level a fifth. The major power over coastal development lies at the federal level with the Corps of Engineers, which reviews most coastal development from the standpoint of overall public interest. A second group includes a number of other federal agencies with specific congressional mandates: concerning environmental protection, transporation, and energy, for example. State agencies deal with coastal policies and other regulatory functions, such as land, wildlife, and water-pollution control. Even local governments exercising police powers have some control over coastal development. Almost all land-use control is exercised at the local level, in fact.

The newest factor is the coastal zone management program resulting from the Coastal Zone Management Act of 1972. It has no power to *control* coastal uses, except limited override power in certain circumstances. Through grants to states, however, it has had a tremendous *influence* in bringing attention to the coastal zone.

What common characteristics do these actors exhibit in reviewing coastal development? Their primary purpose is to resolve *conflicts* aris-

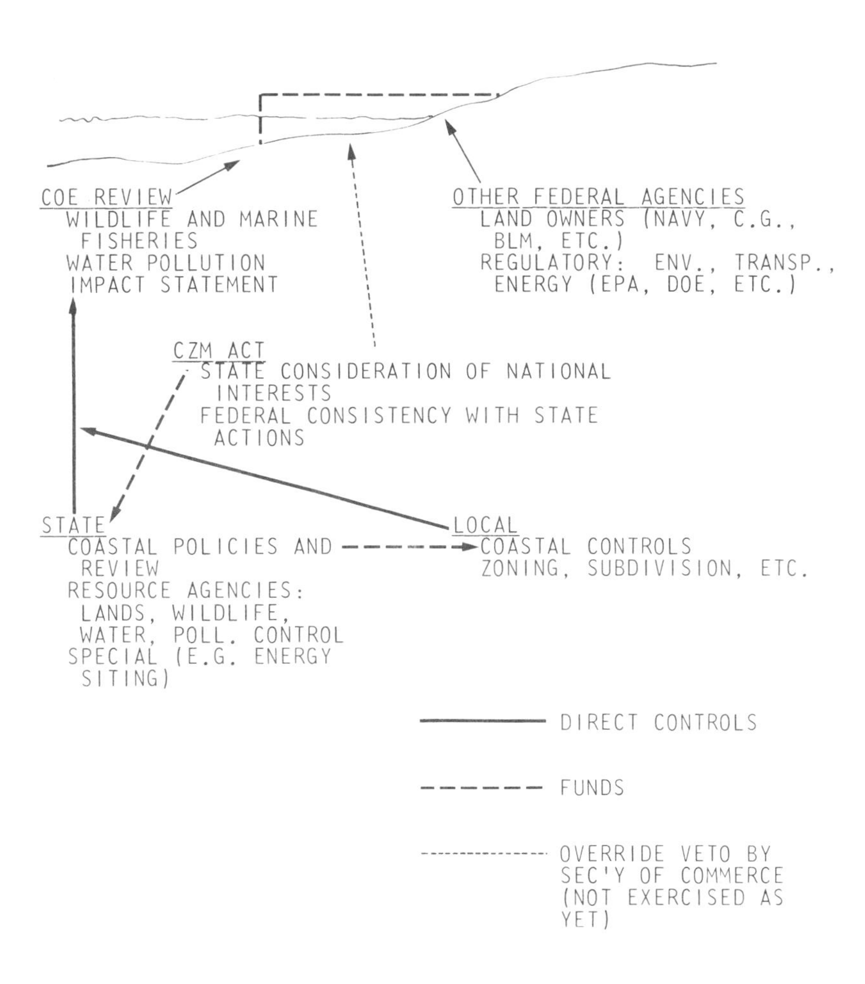

Figure 5.l: Public Agencies Involved in Coastal Siting Decisions

ing over the *impact* of development on coastal environments and on public recreational access to the coast.

Conflict resolution is a key factor in the United States. The Stratton Commission saw the primary coastal problems as those of intense conflict between competing coastal users. Hence we have a lot of stress on multiagency participation and public involvement in CZM. The conflict basis of CZM differs from "increasing resource use." Resource use is the underlying basis for other resource management activities—for example, fisheries management or mineral development. It is important to keep this distinction in mind.

A second key term of CZM is *impact*. Usually three types of impact are of concern in CZM: environmental impacts, which deal both with the biological and aesthetic side; public or recreational-use impacts; and community-character impacts, which are those underlying the growth management movement. Normally, impacts are seen as negative, something to be ameliorated, softened, or reduced. "Negative impacts" are measured in the sciences of coastal oceanography and geomorphology. These sciences describe how the natural processes work, and show the degree to which an impact deducts from the natural process. Another measure of impacts is in the historic doctrine of public trust. Impacts are seen as deducting from the right of citizens to have access to these resources. Because of these natural-system and public-trust values, the nature and purpose of a use is less important than the severity of the potential impact. Any type of coastal use could cause an impact, so CZM efforts have to include all uses within it.

CHARACTERISTICS OF CZM

Given these two fundamental notions for CZM, we can discuss six other characteristics. First there is the focus on *development*. CZM focuses on major visible changes in the use of land or water. This often involves the siting of new facilities in previously undeveloped areas, or in areas where the use was not common before. This characteristic arises out of the concept of impact. Major development arouses the greatest fear or likelihood of impacts, and thus is a primary area of concern.

The second characteristic is *comprehensiveness*. Any use causing an impact singularly or cumulatively is important, since any use could affect the environment or public rights. Thus, under the Act, all land and water uses affecting coastal waters are to be managed.

The third characteristic is that CZM primarily relies on *policy* statements to help decision-making, not plans. Policies normally refer to the way in which environmental-impact and public-use considerations are to be part of the review process. The policies usually are very general and allow little predictibility of outcome. Precise standards, as in the field of pollution control, are very rare. Also, allocation of space, as in an ideal zoning program, is rare in CZM programs. If done at all it is at a very general level, such as dividing the cost into broad categories like natural, conservancy, or developable.

A fourth important characteristic of CZM is seeking *balance*. Since the impact of development is a problem, not the nature of the development itself, CZM requires a balance of development proposed with the need for offsetting environmental and access impacts. In resource-management programs such as fisheries or minerals, where optimal use is a primary goal, balance is really not needed. Many environmental and access problems originate because of lack of balance in other resource management programs.

The next characteristic of CZM is its focus on project *review*. CZM reviews projects proposed by others. A permit system normally results. Planning for the most appropriate use in this particular area is not the primary concern. Projects are normally reviewed only once, when initially proposed, or when major changes are proposed. This characteristic arises from the conflict roots of CZM. The project-review stage is the most visible, and the time when conflicts are intense.

The sixth characteristic is the involvement of all agencies, interest groups and the public. Again this arises from the conflict-resolution root of CZM.

WHAT CZM IS NOT

We have discussed what CZM is. What is it not?

CZM is not a planning activity in the traditional sense of that term. Traditionally, planning identifies desired future uses in an area, and marshals public resources to help bring the uses about. Little attention is given to specific goals or needs of particular users or sectors of society. There is little long-term perspective, or vision of the future. You rarely see a conception of how a coastal area should look in ten, twenty, fifty years. The "planning" which does occur attempts to develop policies and procedures to aid in direct resolution of a contemporary conflict. A broader group of citizens and future generations are left out. CZM does not plan because there are really no resources. Therefore, planning may be futile. CZM is at the mercy of developers. It is the developers who have a vision of the future (often one with which many people disagree) and all CZM is doing is slightly modifying the vision that the developer imposes.

CZM also is not a resource-management activity. The resource is not carefully defined, nor is their control over it as in such things as port management, business management, or fisheries management. Resource managers know a lot about their resource, whether it be fish, cargo, water, drainage or minerals, and have experts who know how the resource relates to the environment in which it is found, how it relates to the economy, about technology for using and producing it, and about trends in its use. The resource manager's goal is clear: to optimize the use of the resource and sustain it for future generations. Resource managers also market their resources, creating constituencies for their use.

CZM also should be distinguished from specific regulatory agencies of government like the EPA, OSHA,* Federal Aviation Administration, Nuclear Regulatory Commission, which oversee specific aspects of the technology, transportation or economic dimensions of a particular industry. These agencies tend to deal with the internal and day-to-day workings of an industry; CZM is not doing that.

Nor is CZM a service agency. Except with one minor exception, no real subsidies are administered. There is little research or technical-data assistance, as one would find in service agencies like the Agriculture Department, Soil Conservation Service, or Maritime Adminstration.

What CZM is most like is zoning. Development projects are exposed to general review to determine who is concerned with what types of impact. The greater the change in land use, the more scrutiny involved in the review process. This is very similar to a change-in-zoning process, which often occurs at the local level. In fact, a classification system similar to that found in zoning is beginning to find counterparts in CZM. Procedures for expediting minor projects are beginning to come about. CZM has also made a major contribution beyond what local zoning has done: appealed to the state government to get a broader review of a proposed development project. This rarely occurs in zoning.

CZM ACTIVITIES

We have characterized Coastal Zone Management. How do we assess its activity to date? What have we gained from a program geared to resolving conflicts and ameliorating impacts?

Certainly a great deal of public attention has been given to the coast over the past decade. The focus is definitely there. It has become an area of concern to a number of people. Forums have been created. Debate is essential for clarifying interests; clarification means future solutions may be possible. Policy terms are becoming more specific. The idea of mitigation—a very fuzzy term three or four years ago—is certainly clearer today, even though more work is needed. The idea of water dependency in CZM, and other conceptual tools, are emerging. We certainly have better protection of wetlands, for example, and we are thinking more about

*Environmental Protection Agency Occupational Safety Hazards Acts.

dealing with coastal hazards. We are incorporating the notion of how nature works into the engineer's solutions for coastal problems. We have gained some very substantial innovations in intergovernmental relations, one of the major CZM achievements.

ADDITIONAL CZM TASKS

Should we get more from CZM than impact amelioration and conflict resolution? I think we should. Impact amelioration may only postpone conflict to a later date. Also, if coordination and negotiation among all interested parties is the primary process of CZM, and the common denominator and "pre-dividing" are the outcomes, this could result in a proliferation of the status quo. Interest groups will use CZM to solidify their positions; new ideas will have difficulty entering the system. The total population should be considered in CZM, not just adversaries in a particular conflict. One way out of this impact-conflict framework is to ask some broader questions: What are the best uses of the coastal zone? For particular coastal environments? Can environmental and recreational enhancement be *planned for,* instead of negative impacts on those resources being *protected against?* And if the coastal zone is partly owned by the public at large, how can the public be made more a part of it?

There is so much potential to what we can learn and do in the coastal zone. A broader conception of CZM is needed. The Federal Office of Coastal Zone Management has been developing a broader conception of Coastal Zone Management, despite a Congressional mandate emphasizing the impact-conflict model. We can attribute this broader conception to energetic and positive leadership on their part. For example, the urban-waterfronts program is encouraging restoration for public access. The fisheries-assistance program is trying to enhance fisheries development in the coastal zone.

Some other innovations are creating real resources for CZM. A coastal conservancy has been created in California with the power to acquire land, hold it, and use it in certain ways. An estuarine sanctuary program has been designed to increase long-term research and public awareness of coastal ecosystems. A third innovation is the reuse of urban waterfront lands for recreation and water-related commerce. A fourth is creative use of dredged material to create new environments.

Some other ideas should also be considered. We should begin to market both access to and knowledge of the coastal zone, through shoreline exhibits, interpretive centers, historical signs, walking tours and waterfront parks. We should build closer ties between CZM and those who manage submerged lands, leading to enhanced aquaculture development and marine parks.

There are potential problems in suggesting CZM have real

resources. CZM, like other resource management programs, could get captured by special interest groups. CZM got started because of narrow, special-interest perspectives of natural-resource management. But positive programs which complement and develop the initial intentions of CZM are appropriate. If we focused attention on enhancing public access, environmental restoration, and access through positive programs, we would have a real resource with which to complement the basic purposes of CZM.

A second area needing future attention is the development of new and broader *information* sources for CZM. We have been influenced by coastal ecologists and geomorphologists, economists and lawyers. We should actively pursue other sources as well. *Historical* studies for example. The coast has always been managed in some way or another. A project just finished in Seattle analyzes the period 1850–1930, when the primary goal was to fill wetlands as rapidly as possible for industrial development. The author characterizes fill activities during that period as a form of coastal management. It is completely out of phase with what we would call a coastal management goal today. We have yet to think how resources used in the past might give us ideas for today and the future. Another area of study to broaden our perspective is coastal *culture.* Change may occur for reasons totally unrelated to law or economics. Attitudes toward growth, land, and other resources may be powerful CZM forces, yet all of our attention has been put on economic incentives and legal mechanisms. Finally, *transnational* perspectives have yet to be explored by coastal managers.

There was a reason why we got into CZM in the first place. We recognized the coastal zone as an extremely valuable place, important to many individuals and organizations, and the scene of some very intensive conflicts. In the last five to ten years we have done an excellent job identifying the coastal environments and amenities, and the impacts that affect the coastal zone. And we have organized arenas in which to debate and resolve some of these conflicts. But we have to capitalize on what is important and unique to the coastal zone. We must deliver our message to a much broader group, describing how they can use and understand their resource to enrich their own lives.

CHAPTER 6

Coastal Management in England and Wales

CHRISTOPHER WAITE

Planning Department
Kent County Council
United Kingdom

DEVELOPMENT OF COASTAL POLICY

Although some prior attempts had been made, in England and Wales a national planning framework with sound town and country planning principles derives from the 1947 Town and Country Planning Act. For some twenty years the operation of planning became increasingly sophisticated before attention was focused on the coast in the 1960s. Recognizing the unnecessary intrusion upon the undeveloped coast by built development (urban development in ribbon form, major oil refineries and power stations, holiday development, including large caravan parks) the government issued a Circular from the Department of the Environment in 1966, urging County Councils to define Coast Protection Areas and tighten development controls.

This initiative to conserve as much unspoiled coast as possible was built on sound foundations:

> *Coastal Variety.* English and Welsh coasts range from dramatic cliffs and headlands—such as the White Cliffs of Dover or Lands Ends—to flat estuarine marshes—such as The Wash—or unique spits or peninsulas—such as Chesil Beach or Dungeness.
>
> *Undeveloped Coast.* Seventy-five percent of the coastline remains undeveloped, with eight hundred miles of outstanding scenery, rich in historic associations and natural conservation.
>
> *Footpath Access to the Coast.* Historically footpaths, bridleways, and droveways have established a complex network of paths across the countryside and to the coast. These routes, originally for working purposes, are

now maintained as public paths and are increasingly, if not exclusively, used for recreation. Accordingly, although much of the coastal marshes or cliffs is in private ownership, the public has right of way across these lands, provided they keep to the recognised paths. Similarly, access along the beach or foreshore—and associated activities like shell fishing—is recognised as a de facto right on many private coastlines, or may even be subject to ancient rights of charter.

Public Ownership of the Coast. Substantial lengths of the coast are owned by public authorities. Those in the District authorities' ownership are most likely open for recreation and general access. Water authorities may also own the sea wall, where one is necessary, which can provide an additional linear access along the coast. (Contrarily, large coastal areas held by the Ministry of Defence for training or firing range purposes deny public access even to former footpaths, but these can be considered reservoirs of wildlife and archaeological interest, relatively undisturbed by modern agriculture.)

Other Protective Ownerships. In the 1960s the National Trust owned one hundred fifty miles of coastline. The Nature Conservancy Council owns substantial nature reserves and has designated five hundred miles of coast sites of special scientific interest. This latter designation requires that the Conservancy be consulted on any development applications submitted so that it can draw attention to wildlife interest.

The authorities responded to this conservation initiative rather over-enthusiastically, largely by expressing negative control policies and excluding development except for limited categories. Coastal Protection Areas extended well inland—often to wherever the coast was visible—rather than where coastal policy was most applicable. Coastal policies became an extra overlay on top of other countryside policies, or policies for rural settlements, which stood or fell on their own merits. The Coastal Protection Policies were thus somewhat discredited: they applied over too large an area; duplicated other policies; were negative and inflexible in concept. The 1960 Circular invited attention to new policies as a matter of some urgency, which led to their ad hoc adoption outside the normal plan-making process, leading in turn to a lack of policy integration.

The Countryside Commission wished to go further, by staffing special joint committees of County and District authorities—with block-grant funds from central government to ensure a coordinated approach, positive proposals, and negative controls—but this was not achieved. It could have taken lengths of coast out of the planning mainstream, not giving due recognition to their interrelationship with the hinterland. Instead, in 1972 the Department of the Environment issued a further Circular, asking County authorities to designate as "Heritage Coast" those lengths identified nationally as outstanding. The County structure plans should recognise the need for long-term conservation and management in these coasts. Local authorities retain the powers to prepare and implement management plans, which the Circular also encouraged as a priority. The Commission retains the influence to encourage management by offering grant aid for projects. Nearly all eight hundred miles of the forty-two Heritage Coasts—forty percent of the undeveloped coast—have now been so designated in County structure plans. Additionally, the National Trust undertook "Enterprise Neptune," a national appeal to bring further outstanding coastline under protection. This has been a notable success; the Trust now manages four hundred miles of coast and is campaigning for "the next one hundred miles."

To illustrate how this policy approach has grown, we may examine the experience in the County of Kent, which has a one hundred seventy mile coastline, thirty-seven percent of which is built up. Kent County's initial policy plan, in 1966, adopted a one-mile-wide coastal belt to protect as a convention before a full survey could be carried out. In 1967, Coastal Preservation Areas were defined. These varied in coastal depth from three hundred yards to five miles, and were broadly based upon the land-area visible from the coast, i.e., the area in which development would have an impact. Strong policies were adopted for the undeveloped coast, to preserve unspoiled natural amenities. Three coastal policies were derived to permit a sliding scale of coastal access, based on the scientific value of the undeveloped coast, allied to its landscape quality. This approach resulted in over-large preservation areas and policies too severe in restricting development, including desirable forms of recreational access and management.

Accordingly, the approach in the 1977 plan submitted to the Secretary of State for approval identifies coast and countryside zones broadly by physiographic characteristics, land use, ecological character and value, and the like. Policies can be better related to a particular stretch of undeveloped coast, and integrated with policies for the countryside or for built-up areas adjoining. Areas of special significance for agriculture, landscape, nature conservation and coastal scenery were identified, including extensive coastal areas. Policies apply to each of these areas, and a local subject plan is now being prepared (by the County Council) to define the areas upon an ordnance map, and to interpret policies, including those for a conflict of interest between, say, improved agricultural production and the conservation of wildlife or landscape. Unspoiled coastal areas not subject to these particular policies are protected by a policy which states that development detracting from scenic quality or scientific value will not be permitted, even in the adjoining countryside. A special case of overriding need must be made before the policy will be set aside.

UNRESOLVED ISSUES IN COASTAL MANAGEMENT

Coastal settlements and the undeveloped coast share many problems with their inland counterparts, but have additional characteristics unique to the coast, like a limited hinterland, "one-sided" transport systems, coastal erosion, and threats from the sea. For coastal management to succeed, it needs to grapple with these characteristics. The following four issues are particularly significant in the United Kingdom:

Agriculture and Forestry These activities have modelled the landscape we now prize, but are almost entirely exempt from planning control (except for large farm buildings and farm dwellings). The landscape can be drastically altered by a change in farming practice to meet economic circumstances or reflect modernization. Much grazing land has come under the plough in recent years, including traditional downland, grassland, and marshland never cultivated before. Modern farm machinery, and the wish to avoid unproductive field corners, has led to much larger fields—without hedgerows and trees. Alarming changes are taking place in parts of the country, resulting in loss of landscape features of historical or natural value. The scenic interest of the undeveloped coast is being depleted by the loss of ancient earthworks or field systems, woodlands, semi-natural grasslands and the like, and footpaths now cross large, ploughed, barren fields where they might formerly have followed a hedge.

Recreation and Tourism. Increases in leisure time, mobility, and affluence have led to greater demands for coastal recreation and tourism. Currently visitors from abroad, particularly Europe, are holidaying in England and Wales, due to monetary exchange rates as much as the attractions of the country or responses to marketing. At the same time tourist requirements are changing, particularly toward self-accommodating mobile holidays rather than traditional stays in one seaside resort. Day visitors from urban areas add appreciably to crowding in the peak summer season. The recreation industry is an outstanding economic opportunity for many coastal towns, but it has created a climate of decline in resorts less able to respond to the changing market. Growing numbers of holiday-makers have led to excessive demands for holiday accommodations like caravan parks upon some stretches of coast, or to erosion of habitat.

Harbor/Port Development and Power Stations. Some statutory bodies are exempt from normal planning control when undertaking development for their own purposes, on land owned for that purpose. Consultation with local authorities is necessary, as is consent from the government department concerned, e.g., the Ministry of Power for a new electricity generating-station. These developments may well take place contrary to the wishes of the two tiers of the local planning authorities. In effect the decisions are made by central government, and local authorities are poorly equipped to assess the need for the development in a national/regional context, which should be set against the local/county impact. Consequences of course extend well beyond the site in question, not only in a visual sense in cases of large power stations and overhead power lines, but, for example, in re-siting hovercraft services or developing roll-on roll-off freight services in Dover Harbor. In this instance the transport pattern in the town—the major passenger port in the country—was changed, affecting residential and commercial areas previously free from the noise of hovercraft or the discharge of heavy articulated container lorries from the new "ro-ro" berths.

Two-tier Authority Implementation Gap. Objectives and policies for the coast set out in the County structure plan may be somewhat diagrammatic. Detailed policies and applications are the function of a local plan, normally the responsibility of a District council. If a District council lacks financial or staff resources to prepare all the appropriate local plans for its area, it may well give low priority to coastal management. The County authority could prepare the local plan if invited to do so – or in default of progress – but these options are seldom exercised. Accordingly, there may be a plan-making or policy-implementation gap even for a Heritage Coast area. This gap more likely results from lack of resources, rather than a dispute in priorities between the two tiers of authorities.

COASTAL MANAGEMENT – A WAY FORWARD?

England and Wales have a long experience of plan-making and policy-formulation; the tools are now fairly well-tried and sophisticated. Policy objectives to balance necessary development at the coast – largely in or related to a developed frontage – with the conservation of the best of our unspoiled coastline, are well-established and generally accepted. Problems remain, but can be lessened or overcome by cooperation:

At the international and national level. Dialogues which have no regard for man-made political boundaries are being undertaken in western Europe, including an International Seminar in Kent held in May 1979, attended by maritime authorities from Holland, Belgium, northwest France and southeast England, which was the second of a series on "Management of the Littoral Channel and southern North Sea: The Next Ten Years." The opportunity to attend a Conference on Comparative Marine Policy at Rhode Island is similarly welcome.

At the regional and local level. Joint teams and finance commit participating authorities to meet goals. In my county, for example, all local plans have a joint-team membership: that is, if a District is responsible for local-plan preparation and adoption, a County Officer is a member of both the working team and the steering group of Chief Officers, and vice versa.

Between government agencies and local authorities. Ironically, the coast is threatened by intrusive development from one arm of authority probably more than from the private sector, while another arm is charged with conservation. Local authorities need to coordinate plans, recognizing that sites are needed for a realistic national/regional program. Government agencies need to fully justify their development proposals, opening up to public debate the criteria for site selection and the methodology used for assessing need. Other agencies which grant aid to private owners – for example the Tourist Boards, Ministry of Agriculture, or Forestry Commission – should seek to ensure balanced objectives.

Between the private sector and public authorities. Partnership schemes may be most successful where the authority can assist financially in land management in exchange for public access. Or where the private sector can inject capital and technical know-how, but needs support and assistance from local authority for public-relation purposes or practical problems for, say, off site works.

What is much more lacking in England and Wales is practical implementation. Cooperation is desirable, if not essential, at the macroscale. But coastal management can best be generated from a small start, building up goodwill and examples of achievement which will encourage enthusiasm to initiate further schemes elsewhere. Management for limited areas has a number of advantages:

> It is a direct application of the policy objective. It is positive and not negative.
>
> Works can be small-scale and low-cost, but bring immediate benefits. At a time of competing priorities for limiting funds, this can secure support for management schemes.
>
> .Works can be of direct gain to the landowner as well as the visitor. This is particularly important, for the relationship with the farmer/landowner is critical.
>
> The works are often simple enough and local enough for volunteer labor, including school children, to participate. This is of great educational value in assisting (young) people to understand the natural environment as a living, changing scene, and to care for it.

Pilot management projects have been established on these lines in Glamorgan, Dorset, and Suffolk, with Heritage Coast Officers financed at least for the first three years by the Countryside Commission and grant aid towards the work involved. A larger number of similar schemes are being undertaken by local authorities, with a ranger or warden appointed to lead. These schemes also receive grant aid from the Commission. These are small beginnings. The management process involves different skills, not least the social skills of relating sympathetically to various interested parties, enabling advice and actions to be requested, providing what is seen as aid rather than control. To act quickly and to avoid the reality or image of bureaucracy, the project leader needs to be given substantial delegation, including a budget he may call upon for small expenditure without recourse to a steering committee.

The way forward is not only through acquisition of new skills and techniques, although these are necessary. It is as much a matter of political will and active public involvement.

LEGISLATIVE AND INSTITUTIONAL BACKGROUND

The town and country planning system in England and Wales is established under a series of Acts of Parliament issued from the Westminster Central Government. These Acts establish the authorities responsible for planning and the duties and powers of those authorities. Statutory Instruments issued by the Department of the Environment

(DOE), headed by a Secretary of State with a seat in the Cabinet, add details for powers and procedures to carry out relevant duties. The Acts and Statutory Instruments are mandatory. Further statements of policy and guidelines are issued as government White Papers or planning Circulars from the DOE. Since 1974 the responsibility for planning essentially lies within a two-tier local government system—County Councils and District Councils:

> *County Councils.* A shire county (metropolitan counties are not treated here) typically extends over some one million acres of rural countryside, with a population of up to one-and-one-half million people living largely in towns of from twenty five thousand to one hundred thousand persons or urban areas up to three hundred thousand persons. A maritime county might have a coastline of some twenty to fifty miles.
>
> *District Councils.* These smaller areas within counties—perhaps ten to a shire county—may have populations of one hundred thousand, of which perhaps half will be in towns centrally situated in each district. A maritime district might have a coastline up to ten miles in length.

The duties and powers of these two tiers of local government consist of three elements: planmaking; planning control; and implementation. These duties and powers are shared as follows:

Plan Making

County Councils are charged to prepare a strategic plan for the whole county, setting out and justifying objectives and policies for fifteen years forward. The comprehensive "structure plan" will include policies for the development of industry, commerce, housing and transportation, and for the conservation of the environment. It will concern itself with issues of strategic importance, indicating where development or conservation measures will be undertaken. However, it will not precisely define sites for development, or detail control policies for local plans. The structure plan is to be monitored and reviewed when appropriate (perhaps in three year cycles).

District Councils are to prepare local plans for all or part of their district. These plans must conform to the fundamental policies of the structure plan. They will expand upon it at a local level, proposing a program to implement policy in the knowledge that resources are available. Local plans may be comprehensive plans for whole districts, subject plans for particular issues (such as coastal management plans), or action area plans (such as environmental improvement plans for run-down housing areas). Local plans are also to be monitored and reviewed when appropriate.

Planning Control

District Councils exercise development control; all land-use applications are submitted to District Councils. The District Council may permit any application which conforms to the fundamental provisions of the structure plan or which is not inconsistent with a local plan. Any application which it wishes to permit and which does not meet these criteria is referred to the County Council. A District Council may refuse any application, setting out grounds of refusal, and may also consult the County Council.

County Councils only determine particular classes of applications: mineral extraction; County Council development after consulting the District Council; or applications referred to it by a District Council.

Implementation

County Councils are responsible for education, social services, highways, and waste disposal, all of which lead to substantial development proposals. They also share responsibility for a number of other facilities, including informal recreation, sports development, and the conservation of landscape or wildlife.

District Councils are responsible for housing, environmental health, and waste collection; they also share the conservation and recreation functions set out above.

This brief summary inevitably oversimplifies the functions of the two-tier authorities. In fact the divisions are not nearly so clean-cut. Close cooperation between the two tiers is required, or overlapping and duplication result. In practice the County Council may prepare local plans which have significant strategic elements or which extend over more than one district. Similarly, proposals at a local level may be promoted by a County Council if the District Council is amenable. County structure plans are submitted to and approved by the Secretary of State after public participation, including public inquiry before a panel appointed by the Secretary of State. Modifications to the plan may be made by the Secretary of State. Local plans are adopted by the authority—normally the District Council—which prepared the plan. This follows public participation and inquiry before an Inspector, appointed by the Secretary of State, who reports suggestions for modifications to the promoting authority. In each case the County Council must certify that the local plan conforms to the structure plan.

Applicants who receive refusal notices may appeal. The Secretary of State is empowered to appoint an Inspector to hear both sides. Interested third parties are also heard at the inquiry and are able to cross-examine witnesses. The Secretary of State decides upon the appeal, based upon his Inspector's report. But there is no appeal system open to the public if a development is permitted by the authority. (This is justified because elected members representing the public have made the decision.) This applies equally to development proposed by and determined by the same authority. The public may only generate new legislation through central government elections, or local policy or by-laws through local elections. There is no means of voting legislation or finance on the lines of Proposition Thirteen in California.

The essence of the two-tier system is for larger authorities to prepare strategic plans approved by the Secretary of State to ensure a coordinated regional/national overview. Local plans are to be determined locally, subject to adequate public participation. Control is also to be exercised locally, subject to conformity with the countywide strategy. An aggrieved individual may appeal to the Secretary of State as final arbiter if he feels he has been harshly treated by a refusal notice or by permission with too-onerous conditions.

Two national government agencies also play a significant part in planning, implementation, and management of the unspoiled coastline:

> *The Nature Conservancy Council* fosters awareness of the role and importance of nature conservation and the need to conserve wildlife. It advises central and local government on consequences of plan making, development control and implementation. The Conservancy, in many instances, is consulted by statutory right, where a site of special scientific interest is threatened. The Conservancy carries out detailed biological research and surveys, and safeguards sites and species by establishing nature reserves.
>
> *The Countryside Commission* is an independent statutory agency operating throughout England and Wales. Their main tasks are to promote conservation and enhancement of landscape beauty; to provide facilities for informal recreation in the countryside; and to provide better public access to rural areas for open-air enjoyment. This promotion is carried out by exhortation to local authorities and the private sector, backed by grant-aid powers through a budget from the DOE. The Commission also seeks to reduce conflict between town and country interests through understanding of and care for the countryside, and through information and interpretive services. Accordingly, the Commission encourages open farm days, publicity leaflets, conservation guidelines, and the like.

Other agencies with significant effects upon conservation of the countryside and coast are voluntary bodies such as the Royal Society for the Protection of Birds, which not only encourages conservation of wild birds but research, promotion, and maintenance or restoration of habitat. It also manages some seventy reserves. The Wildfowl Trust and the County Trusts for Nature Conservation exercise similar roles. Not least, the National Trust, through its "Enterprise Neptune" campaign to buy additional coastal land, is a major owner of many dramatic or remote unspoiled sections. Established under a special Act of Parliament, the Trust raises funds through membership subscriptions, donations, and bequests; it has a membership over 750,000. It does not receive government funds, but concessions for tax are given to private owners in exchange for property bequests. Access is free to members, at a charge to the public at large. Land and property are now only accepted as gifts by the Trust in a condition or with a maintenance fund such that they can be managed in a fitting manner.

A number of statutory authorities also make heavy demands upon the coast or assist in maintenance. The Central Electricity Generating Board constructs and operates power stations frequently sited at the coast for water-supply and cooling puposes, or, in the case of nuclear energy, to gain a location remote from population. Port and harbor authorities operate ports, ferry services, and marinas. Regional water authorities maintain sea walls and drainage patterns. These authorities generally have powers to fulfill their statutory function on operational land without planning control or public consultation/participation. Large developments, such as power stations, are subject to consultation and public inquiries.

In a separate field the English Tourist Board and its regional Tourist Boards promote the expansion of tourism, tourist facilities, and the up-dating of existing accommodations and facilities in resorts to meet the changing tourist market. The Board receives central-government funding, and the regional Boards are also normally funded by local authorities. The Tourist Boards seek to reap economic benefits in a manner which will not materially harm the coast.

At the most local level, beneath District Councils, Parish Councils are also elected for small villages and rural areas. These Councils may raise a rate fund from householders (as do County and District Councils) for local facilities or improvements, such as the maintenance of a village common land, or village green.

CHAPTER 7

Coastal Management in The Netherlands

HENRI WIGGERTS

Delft University of Technology
Netherlands

INTRODUCTION

The Netherlands – or Holland – is one of the smaller independent coastal states in Western Europe. Like Norway, Denmark, Western Germany, Belgium, France and Great Britain, it borders on the North Sea. The Dutch coastline stretches to about one thousand miles, including the coastline of a number of islands lying before the mainland; a rough estimate on the mainland coastline is four hundred miles.

To make this part of Europe more familiar, a comparison in scale can be made:

> The total surface of The Netherlands corresponds to that of the states of Connecticut, Rhode Island and Massachusetts together; the population number is not much higher (about fourteen million).
>
> The Dutch coastline spans a distance comparable to that between New York City and Boston, but estuaries at extreme ends of the coast give entrance to foreign harbors, namely Antwerp in Belgium and Emden in Western Germany. This brings an international dimension to the problems of Dutch coastal management.
>
> More international aspects: Three European rivers – Scheldt, Meuse and the Rhine – conflue in the southern part of Holland. They are important for inland shipping. Of primary importance is the Rhine, forming the main connection between the world seas and the industrial heart of Western Germany, the Ruhr area. This area is about as far up-Rhine as Albany lies up-Hudson.
>
> The most inland part of The Netherlands – and the only hilly country, though at the highest point not more than twelve hundred feet above sea level – is South Limburg. It is not more than one hundred twenty miles from the sea, like the Catskill Mountains just over the Hudson River in New York.

A gulf in scale comparable to Long Island Sound—the Zuidersea—was closed off from the North Sea half a century ago. Great parts of this former gulf have been reclaimed in the meantime, most of it put into agricultural use.

The southern part of The Netherlands shows a deeply indented coastline, not unlike Rhode Island. In reaction to the stormtide of 1953, which flooded great parts of the country and took eighteen hundred lives, the so-called Delta Plan was executed to close a number of estuaries from the sea by earthen dams. This is a several-billion-dollor job because it means mastering the tidal streams in the area where the average tidal differences are about ten feet and where the sandy bottom is easily erodible.

From these comparative remarks a global image of the Dutch coast emerges in four parts. Coming from the southwest, going along this coast to the northeast, and adding some color, this might read as follows:

1. Behind the first stretch of about fifty miles lie the islands and deep estuaries where the Delta Plan is in execution. The Delta Plan has been—and still is—severely disputed, because a number of people feel too many environmental values have been sacrificed to safety against floods. The average tidal difference along this part of the coast gradually goes down from fifteen to eight feet.

2. Behind the next stretch of about eighty miles there is a chain of dunes, with broad beaches, protecting the central part of Holland. Besides being popular holiday resorts for Dutch and German people, these beaches are an important day-recreation facility for the big cities nearby, like Amsterdam, Rotterdam and The Hague. The chain of dunes and beaches is broken only by a few harbor entrances, including those to Rotterdam and Amsterdam. Tidal differences average from five to three feet.

3. The third stretch of about one hundred miles is along the Frisian or Wadden Islands. The gaps between these islands leave the Waddensea—the remaining part of the Zuidersea—in open connection with the North Sea. These gaps cause channels—often of great depth, down to forty feet—in the Waddensea, which run between sand flats. The Waddensea is recognised as having unique biological importance, especially in regard to migratory birds and lower-water organisms. The average tidal difference goes up again, from three feet to seven feet in the north.

4. During the whole trip along this sandy coastline, to the left lies the North Sea. This is a shelf sea of the Atlantic Ocean, bordered by Norway, Denmark, Germany, Holland, Belgium, France and Great Britain. The North Sea is polluted by all these countries: all have shipping lanes, extract sand and gravel, explore for gas and oil, fish, and conduct naval maneuvers. There is—and should be—growing concern for water quality and marine life in the North Sea.

OLD PRIORITIES AND NEW DOUBTS

Without its defense line of dikes and dunes, about two-fifths of The Netherlands would be flooded permanently or twice daily by the working of the tide. More than half the total population lives in this part of the country. The vital need to protect it against the sea is evident, as is the need to keep it dry. Over many centuries the Dutch have come to accept a strong controlling and maintenance regime of dikes, dunes and water management.

That lowest part of "the low countries" was once one vast morass behind a barren dune belt. By successive "poldering" (building a dike around a certain area and pumping it dry), the present situation was attained. This process started about ten centuries ago, which may account for the fact that reclamation of shallow areas like sandbanks and mudflats, until two decades ago, had been another undisputed activity.

Also, for centuries the Dutch have been a seafaring nation. World shipping and trade brought wealth to this small country, that otherwise would have been entirely dependent on Germany. Subsequently harbor entrances, harbor facilities, and connecting transportation, like inland shipping, got high priority.

When in the last century the need for a good drinking-water supply became evident, the dune areas along the coast quite naturally provided the ground-water supply. This development made it necessary to put the dunes under some protective regime.

All these developments, and their implicit priorities, resulted in coastal-zone management by public or semi-public authorities beginning many decades or even centuries ago. Public access to beaches never became a problem, and private developers almost nowhere had a chance.

During the last two or three decades, however, there have been new developments. Population growth and the resulting urbanization brought industrialization in large part to sea harbors, necessitating an enormous expansion in coastal areas. Motorization, increased consumption, and spills gave rise to a growing concern for environmental quality. Erstwhile, undisputed development priorities are now under discussion. The need for a good defense system against flooding is still recognised, but how high must we put the risk percentage? Is it wise to sacrifice a biologically very valuable estuary, like the Eastern Scheldt in the Delta, for a marginal gain in safety? Land reclamation seems to have disappeared almost entirely as a national priority. Further expansion of harbor-related industries has become almost imposible due to public resistance. Ground-water supply from the dunes has already been overdrawn, and must be reconsidered in light of more modern purification techniques for open water. Recreational use of beaches is still all right, but negative effects must be reconsidered in a wider context, as must a number of infrastructural works like entrance roads, parking spaces and pipelines. Nature preservation and pollution control have become important new priorities. So has energy saving, but fuel importation and gas-and-oil exploration and exploitaton nowadays are very suspect.

This all leads to two main conclusions:

1. Coastal management in Holland stems from a long tradition, and is based on a framework that has grown over centuries.

2. However positive this sounds as a starting point, there is reason to look not only at traditional priorities in the perspective of contemporary problems, but also to reconsider management in a less-restricted sense.

In the next section a brief account of the organisational framework will be given, following which essentials of the Dutch land-use control system will be sketched, with a special eye to new developments relevant to coastal management.

TRADITIONAL FRAMEWORK

By now it must be clear that Dutch coastal management, even in a restricted sense, is a part of the general system of defense against the sea, river floods and the water system. The first attempts to set up some sort of authority for this system date back twelve centuries. Gradually there grew a functional management system that succeeded in surviving subsequent government systems. It grew stronger through the feudal system, through the Republic of the United Netherlands, through the puppet-kingdom under emperor Napoleon Bonaparte, and during the present, decentralized, unitary state with its hereditary monarchy.

At the base of this system operates the "waterschap," which is a law-based cooperation of landowners, each holding votes in proportion to his property. These water-managing authorities (eight hundred in number) have their own technical staff, and have authority in their own field (functional and geographical). Under supervision of the provincial governments, they have their own technical staff in a special body called "Provinciale Waterstaat." These in turn are supervised by the chief directory of "Waterstaat" at the national level, assisted by "Rijkswaterstaat" and attached to the Minister of Transportation and Waterstaat. A number of flood-defending dikes and dunes come directly under provincial or national "Waterstaat," while in some cases municipalities have powers comparable to those of the "waterschappen."

It appears that in this system both decisionmakers and technical staff are experts on and devoted to their tasks, while the hierarchy in supervision guarantees that decisions are well considered and taken in full responsibility. Criticisms on the system come down to it being a closed shop, with a limited functional-technical if not technocratic approach that avoids real public participation or trade-off with outside interests.

In the foregoing, the governmental structure of the present Dutch decentralized unitary state has been indicated implicitly. There are three levels: municipality, province, and national government. On each level there is a division between executive and legislative (or controlling) powers. Procedures for administrative justice stipulate that conflicts over plans and most actions from government are not adjudicated in the courts, but are fought in successive steps in the government hierarchy, eventually ending with the Crown.

PHYSICAL MANAGEMENT AND EQUIPMENT

Parallel with acceptance of governmental responsibility over defense against and use of water grew acceptance of government responsibility for planning and controlling development of the whole environment. The Netherlands has a tradition in urban and regional planning with the following key concepts and instruments:

Municipal allocation plans, the only plans binding for citizen and government itself, in the sense that they are (almost) ministerial for decisions on building permits.

Municipal or intermunicipal structure plans.

Provincial regional plans.

National physical-planning policy.

Each of these concepts carries guarantees for survey and research, intergovernmental deliberation (vertically and horizontally), public participation, procedures for decisionmaking and means for plan enforcement. The whole system built by these concepts interlocks in a consistent way: in principle, "lower" plans must comply with "higher" plans. The system has a *regulative* character (the plans *allow* for certain developments and none other, but do not *enforce* them), though recently some amendments have been made to make it more operational in areas like urban renewal.

As can be understood from above conditions, in coastal management the allocation plans have had limited importance. The regional plans, indicating future developments and offering a framework for trade-off in a wider context, have been helpful in a number of cases. The national physical-planning policy has been acting as a framework and may be more helpful in the future because of recently introduced new instruments specially relevant to the scale of problems facing coastal management nowadays. These new instruments are the crucial physical-planning decision and the structure scheme.

The crucial physical-planning decision is a special procedure, some times referred to as a plan for a "national allocation decision," to be used in cases of such importance or scale that a trade-off of interests can not be reasonably conducted at lower levels of government. The crucial physical-planning decision assures intergovernmental deliberation and public participation. It will be used for decisions on main outlines and principles of general interest for the national physical-planning policy, and for decisions on actual policy plans which deviate from established main outlines and principles. This procedure can be either of an independent or of a supplementary character. The supplementary procedure will be applied when the decision concerned is already subject to regulations meeting the general requirements. The supplementary procedure then consists of a series of "physical-planning injections" relating to the general requirements, which—where necessary—are incorporated in the existing procedure. The independent procedure will be applied in cases where the decision concerned is not already subject to regulations meeting the general requirements.[1]

Structure schemes are especially relevant to all sectors of infrastructure, such as traffic and transport, seaports, waterways, airfields, pipelines and electricity supply. Structure schemes are drawn up more or less on the same basic assumptions—like population growth—and follow the same model, making them comparable in principle and allowing a trade-off for different sector claims.

Behind all this lies the basic concept of *double-track decision-making*. It is recognised that land-use control on the one hand and different government sectors on the other have their own rights and rationality. Each of them has its own institutional and organisational framework and over the years each of them has developed its own specific methods for research and execution, policy and decision-making. This is valuable and should be mutually respected. Therefore it is necessary to reach decisions by both tracks of decision-making. Each track is followed to the end, but of course an attempt is made to synchronize the processes and to make intermediary contacts along the line. The structure schemes and the procedure of the crucial physical-planning decisions are main stations in this double-track concept for coordination in government and management.

Environmental-impact assessment had an enormous impact on the planning process in the United States when it became a new requirement for making locational decisions. Use of environmental-impact assessment in Holland has been studied for some time, and it seems probable that within a short time it will be compulsory in Holland, too, though not with the same impact on decision-making. In the meantime research on the possibilities of better environment-based planning and decision-making has gone on, resulting in a landscape ecological survey of the whole of The Netherlands on a scale of 1:200,000. For this purpose, an inventory, typology, and survey were made of certain components of the natural environment to estimate influences of human activities.

[1]"Waterstaat" laws and environmental-hygiene laws often meet "general requirements." The structure scheme "Inland Waterways," for instance, followed the supplementary procedure.

CONCLUSION

The major factors affecting Dutch coastal management have been indicated. To be more specific on the objectives and instruments would necessitate focusing on special problems, aspects or areas, but a first indication of the range of the field to be investigated can be given.

The Delta plan has been amended recently to strike a better balance between environmental values and safety against flooding. At an additional cost of about one billion dollars, movable storm-flood gates will be built into the closure dam of the Eastern Scheldt to ensure a tidal regime in this estuary. A number of people and organizations think this should have been a crucial physical-planning decision (which at the time was not yet possible) and that even an apparent detail like the future instruction manual for this system is of such crucial importance to marine life in the estuary that it should be established by this very procedure.

The coast of the central part of Holland also will be important. In preparation are structure schemes for harbor development, pipelines, and open-air recreation; those for inland waterways, traffic, and transport are in procedure, and that for drinking-water supply will shortly be revised.

For Waddensea, a specific crucial physical-planning decision is now in procedure. It will be a formal base for the Dutch government to get international recognition and cooperation for the whole of the Wadden along the Dutch, the western German and the Danish coast. New institutional and organizational frameworks for this will be established.

And finally the North Sea, for which an interdepartmental committee for policy issues was set up. Its first task has been advising national government on the location of a liquid natural-gas terminal for The Netherlands. Among its other tasks is advising about an artificial North Sea island, for industry that could be dangerous or otherwise unsuitable for mainland location. Another government committee was installed to initiate and coordinate research in marine matters in general; one of its working groups is for planning and managing marine environment.

In many ways we are still at a loss about the concept of coastal management. Does it stand for keeping a stable coastline, or does it mean control of coastal zone development? The first interpretation would lead to a rather technical approach and the second view would place it as a part of land-use planning. From whatever end one starts considering these matters, one ends up at the other. Good coastal management should merge with good land-use planning and control, and a realistic attempt for this is being made in The Netherlands.

Commentaries

ROBERT KNECHT

National Oceanic and Atmospheric Administration
United States

Some circumstances are peculiar to the US coastal zone. The magnitude of our coastal zone problem is probably greater than most other countries. We probably have the largest area of coastal wetlands in the world. The diversity of the US coastal area is impressive as well. Many nations have coastal zones with two or three major characteristics, while ours has virtually all except one. We have estuarine environments, coastal marshlands, wetlands, rocky cliffs in Maine, bluffs in California, the mountain shoreline of the Pacific Northwest, the fiord-like shoreline of Alaska, and the flat, indented shoreline of the barrier islands of the Southeast and the Gulf Coast. What we don't have are desert shorelines such as in Chile, Peru or parts of the coast of Africa and Australia. It may be a little unusual, too, that we have very highly-developed coastal zones on the one hand and near-wilderness on the other. Consider the Illinois shoreline along Lake Michigan, five miles all integrated, all developed, with little flexibility anywhere, compared to the very undeveloped regions in Alaska, the southeastern United States and elsewhere.

Pressure for Coastal Zone Management (CZM) recently has come from the national level, while most existing control of land use is at the local level. CZM in this country began in the mid 1960s, in two ways. First came the coastal wetlands use laws along the Atlantic seaboard, passed in 1963, 1964 and 1965. Next came the San Francisco Bay Conservation Development Commission, which began in the early 1960s and led to coastal management in Oregon, Washington, California and the Great Lakes. Finally, the Federal Government came into the picture in 1972.

Great regional variations mean different needs and values. Certainly the values of inhabitants of the Oregon and California coast differ from those of people along the Mississippi and Alabama coasts. Mississippi and Alabama place great stress on economic issues, increasing opportunities for inhabitants to earn a wage closer to the national average, whereas Oregon and California are more concerned with opening public access, protecting resources and siting necessary development rationally. Economically speaking, we continue to want the coastal zone in this country to do more of everything, which is probably common to other nations as well.

One principal characteristic of the United States CZM program is that it is voluntary. The national government was not ready to make it mandatory in 1972, and probably won't be soon. Our program works on financial incentives, along with the promise that federal actions be consistent with state programs.

There seem to be three ways to move CZM: make it mandatory; provide enough incentives; or identify a crisis and have people act out of fear. CZM works best when a voluntary program is in place with incentives, a mandatory program is in the background, and a crisis is out on the water. Lacking any one of those three, the program lags.

Another characteristic of the US effort is the minimum of mandated national policy goals in present legislation. Very little in terms of purposes or substantive national policy is contained in present legislation.

Marc Hershman emphasized that the national CZM program was largely process-oriented, and I think that is right. A final characteristic of the US effort is that the program is in the hands of the oceans, fishery and atmospheric agency, and not in the hands of a national planning agency or a national Department of Resources. This is an important point we sometimes lose sight of. It means that the US effort, to the extent it will continue to be led from its present organizational home, will tend to concentrate on coastal water management. The fact that we do not have a national energy program or a central planning body is a substantial difference between the US and many European nations, though the coastal problems seem quite similar.

Coastal management has to be seen to have a positive as well as negative side. During its first years in this country it got a negative reputation—primarily out to prevent development, to protect natural resources, to prevent the filling of wetlands, and so on. We have to emphasize the positive side of coastal management. Marc Hershman already mentioned our initiatives in urban waterfronts and coastal fisheries assistance. We are also working with states on a number of demonstration projects to preserve historic resources or deal with transportation problems.

New Jersey, for example, is just initiating a third year of its CZM-funded beach jitney program, to transport people from parking areas along the Garden State Parkway into beaches, relieving pressure on close-in parking lots. Linking management of important water areas to management of shoreland areas is wise. The estuarine sanctuary program, the marine sanctuary program and the coastal management program are all a part of the same office, fitting into this pattern. In addition, the Office of Coastal Zone Management has begun an ocean-resource planning and coordination effort which ties very closely to more near-shore concerns.

Specific coastal management policies should be developed along the four areas that Marc Hershman mentioned. We must examine the protection of specific coastal resources, the guidance and management of coastal development, improved public access to the shoreline for recreational purposes, and improved government decision-making. More specific goals should be incorporated into amended legislation, and future federal funding of state programs ought to be tied to achievement.

J. R. JACKSON, JR.

American Petroleum Institute
United States

My comments on CZM will be parochial and pragmatic, from an industry viewpoint. These basic remarks may reflect the energy industry's policy on a global basis.

We support CZM principles, and believe the CZM Act of 1972, as amended, is a reasonable piece of legislation. The findings and policies declared have great promise. The statute is balanced and provides for both environmental protection and energy development.

Unfortunately, the promise offered in the legislation has not come about. We must reach an accommodation that will further the energy goals of this country, along with environmental protection. A number of people are dedicated opponents of all forms of energy development. The list of places where they have been successful is very long.

On a global basis, we would not want to recommend export of CZM as we know it today. The world may not need it. I am not sure they haven't already had it in many cases. We may learn from them, rather than vice versa.

We are concerned, as you should be, about access to areas for energy development and institutional delays. We need to expedite rather than delay, and all evidence to date indicates that delays are winning the battle.

NORMAN GILROY

Institute for Human Environment
United States

Architects and planners very often provide the glue between specialists, and in many ways play a fairly new role in the process. The Institute for Human Environment attempts to link government, universities, and commercial organizations to identify issues, to bring reasonable debate, and to transfer knowledge on an international scale between countries with similar problems. It is surprising how often a community in the United States or Europe faces a problem and thinks it is the first to face it ever, when someone else has tackled that problem in substantially the same form, albeit in a different cultural, financial, or maybe even political situation.

Frequently, however, policy issues are rather difficult to transfer. Frequently, practical solutions are much easier to transfer. This group must avoid becoming theoretical and talking to itself. One need only look at lawyers who specialize in coastal management and listen to them talk to think this a foreign country without a dictionary. It could happen to marine management. It has already happened to some degree to coastal management, yet to succeed coastal management really has to work for people and their communities, as well as for environmental balance now and for future generations.

The future-generations aspect is lacking. Policy needs be where action takes place; if there is no thought-out method to allow policy to find its way down to the ground, we are in trouble. We have been in trouble with a great many ill-defined policies going through the interpretation process. We have come in the last twenty-five to fifty years from a concept where man was environmental conqueror to a point where we by necessity must learn to live harmoniously with the environment. The change of attitude is interesting, yet "management" really in many cases has man's conquering the environment at its base.

Now we are starting to see places where instead of engineering solutions a whole range of issues, many of them environmental, is considered. Crisis is still a major generator for change in our society, yet in many ways crisis is more fitted to a conquest approach then it is to a harmonious approach. Perhaps we need to look at different ways to go about creating change. CZM has created crisis in a number of cases simply by forcing the issue of planning, throwing a number of industrial concerns into a great tizzy—yet change has taken place in some attitudes. The Corps of Engineers is a wonderful example.

For effective implementation we need a political climate of acceptance. In some societies that means changing the government's mind. Developing nations have a very different set of requirements. Changing a monarch's mind may need to be arranged. In our society changes in citizen attitudes are necessary. It is slower. Political change usually follows behind it, but nevertheless it is happening.

There are two ways to go about policy formulation in relation to coastal management. One is to stand on the beach and look toward land and all your plans. What is crushing behind you, you forget about. You are carried away by what you see out there when you look below the water. You have a desire to fall below it to analyze it, but you forget about what is behind you. The other is to stand and look up the coast to see the water on one side of you, recognizing them as interrelated and also to begin to see that delicate intertidal zone where they meet.

In the US we have had an extensive program stimulating coastal planning over the last seven years. It made sense in its own terms and became a lubricant for states to begin planning efforts. California has tried for a long, long time to get land-use planning on any kind of basis beyond local. Regional agencies end up with no teeth and slowly disintegrate. The coastal effort in California is the first strong activity to unify planning in one zone of the state, the first time really on that scale. CZM is not planning, but it has set up a planning process in many locations. It is not resource management, and yet it has been the catalyst for innumberable resource management projects. It is not regulatory, but it has initiated many coastal regulations or planning programs. It is not service, and yet service activities have appeared as a result.

The interests of the Office of Coastal Zone Management are now being extended for Marine Policy. That is a very fortuitous circumstance. The national coastal-planning process has been a consciousness-raiser. Public attention has been drawn to an important resource on the coast, sometimes by crisis, sometimes by policy. The next big problem is going to be how to implement planning going on now. How do you come down from national, ill-defined policy to applied state policy or local plans, and then get them to work, given all the political restraints that go on there? That is one place where we can learn from others.

Europe has had a highly structured planning process for many years yet only now tentative movements are being made toward a unified approach to planning country to country. Holland has international ports for oil, containers, and the like, which serve half of Europe, one way or another. A glass of water in Holland has been through at least six countries beforehand. When some particular crisis comes up, such as control of the Rhine water moving through Holland, things obviously don't work. Holland is working. The United Kingdom has the delightful and not-too-profitable opportunity to be an island state, able to turn its back on other countries and never really concern itself about whether its actions affect what goes on across the English Channel. Yet now with changes in the EEC,* the monetary situation, and the interdependence showing up on those levels, small things are starting to happen between different countries which share bodies of water. The Law of the Sea activity, primarily a legal negotiation with overtones from other disciplines, has fallen into chaos. Yet there is interesting activity in the Mediterranean, where the Blue Plan and the Action Program for the Mediterranean show promise.

*Europeon Economic Community

FOUR

Offshore Oil And Gas Policy

Thomas A. Grigalunas
University of Rhode Island

CHAPTER 8

Outer Continental Shelf Oil and Gas Policy: United States

MARTIN H. BELSKY

National Oceanic and Atmospheric Administration
United States

INTRODUCTION

The United States recently reviewed its policies and practices as to offshore oil and gas development and adopted comprehensive reforms to its management and leasing program—the Outer Continental Shelf Lands Act Amendments of 1978.[1] This new law, the result of a four-year effort to reach agreement in Congress and with the Executive branch, was the first revision of the United States oil and gas laws in twenty-five years. It was the product of numerous compromises reflecting the concerns of environmentalists, oil and gas companies, state and local governments, citizen groups, federal agencies, and foreign and international bodies.

Estimates by public and private research units indicate that the United States may have between forty and sixty billion barrels of oil and between 250 and 400 trillion cubic feet of natural gas still undiscovered but recoverable by present technology.[2]

Despite these great potentials, only about three percent of the US continental shelf and margin areas have been leased for development. From this three percent, the United States obtains about eighteen percent of its oil production (less than ten percent of its oil) and fifteen percent of its natural gas production.

The explanation for this delay is complex. However, most concede that it was the result of a variety of interests all raising concerns about offshore activities, leasing policies, safety, and effects.[3]

It is hoped that enactment of the new offshore leasing law will end the "uncertainty and controversy over the rate and location of OCS leasing and development," answer the concerns of "states and localities (which) have sought a greater voice in federal OCS management," and thus expedite balanced and orderly development.[4]

HISTORY

At the end of the 19th Century, oil derricks began moving out into the ocean from wooden piers on the Pacific Coast, but it was not until 1938 that the first platform and drilling well was located unattached to land, in the Gulf of Mexico. By 1946, nine wells had been drilled in the very shallow waters off Louisiana and Texas.[5]

As early as the 1930s, the Federal Government sought to establish its rights to US offshore resources both as against foreign governments and the states. Specifically, President Roosevelt sought legal authority from his Interior, State and Justice Departments to insure "federal jurisdiction...as far out into the ocean as it is mechanically possible to drill wells."[6]

After a series of studies in the 1940s, prompted by the war-time desires to assure increased oil and gas supplies,[7] the Roosevelt Administration and then the Truman Administration embarked on a two-pronged attack to clarify and protect United States rights. It brought suits against California, Louisiana and Texas to establish federal title to the resources of the Continental Shelf adjacent to those states,[8] and also declared the mineral resources of the subsoil and seabed of the Continental Shelf off the United States as subject to United States *exclusive* jurisdiction and control.[9]

Eventually, an international treaty recognized the rights of all nations to resources off their coasts as far out as these rights were exploitable.[10] In the United States, the affected coastal states were precluded from leasing rights off their shores by the Supreme Court in 1950.[11]

The states, however, continued their battle in the political arena—seeking Congressional action to give them dominion—popularly referred to as the "Tidelands Controversy." After the election of a Republican President in 1952, who had pledged in his campaign to assist these states,[12] and with a general recognition that some legislation was needed to authorize the leasing and supervision of offshore mineral exploitation,[13] Congress in 1953 enacted two laws defining the respective rights of the state and Federal Governments and establishing broad policy for federal offshore activity—the Submerged Lands Act providing for state rights in the coastal zone[14] and the Outer Continental Shelf Lands Act providing for federal rights beyond that zone, now called the Outer Continental Shelf (OCS).[15]

OCS leasing proceeded slowly during the 1950s and 1960s. Between the passage of the 1953 Act and 1968, the Interior Department conducted 23 sales, leasing 1,717 tracts, covering 6,411,626 acres. Impact was localized to the Gulf of Mexico and the Southern California coast off of the Santa Barbara Channel. With relatively easy availability of domestic onshore and foreign imported oil and gas, there was little academic, public or political scrutiny of OCS activities or their value.[16]

The lack of awareness of the OCS was sharply altered by a massive blowout at an OCS facility in the Santa Barbara Channel on January 28, 1969.[17] The beaches and harbors of the Santa Barbara coast were heavily polluted and nationwide publicity resulted. In the two years following this blowout, three more blowouts and one fire occurred—continuing to focus public attention on the OCS.[18]

In the early 1970s, a different form of attention was focused on OCS resources. Concerned about decreasing onshore energy supplies and increasing dependence on foreign imports, on June 7, 1971, President Nixon ordered the Secretary of Interior to increase oil and gas leasing and to publish a comprehensive five-year lease schedule. Environmental litigation, however, delayed immediate increased OCS leasing;[19] and in April of 1973, President Nixon again directed the Secretary of Interior to increase OCS leasing—in fact to triple it—from one million acres in 1973 to three million in 1974. The Arab Oil Boycott in October 1973 confirmed fears about potential foreign blackmail and renewed interest in accelerating OCS leasing. In January of 1974, President Nixon instructed the Department of the Interior to accelerate leasing to ten million acres a year in 1975—almost as much as was leased during the previous 20-year history of OCS leasing. This leasing was to be not only on the Gulf Coast but also in previously undeveloped or "frontier" areas on the Atlantic and Pacific Coasts and off Alaska.[20]

Beginning in 1973, numerous Congressional committees in both the Senate and House investigated OCS activities.

The first major Congressional action to amend the OCS law occurred in 1974 during the second session of the 93rd Congress. The Senate, in fact, passed a bill but no action was taken by the House of Representatives.

In 1976, a joint House-Senate bill was proposed but because of opposition by some elements of the oil and gas industry, and the Ford Administration, the bill was not passed. Finally, in 1978, with the support of the Carter Administration, a new bill—S.9—was overwhelmingly approved by both the House and Senate and then signed by the President.[21]

MANAGEMENT AUTHORITY

Responsibility for management of the offshore oil and gas resources of the United States is presently divided between the state, and the federal governments.

The Submerged Lands Act of 1953 gives most coastal states exclusive rights to these resources up to three geographical miles from the coast, and allows certain states bordering the Gulf of Mexico to attempt to prove entitlement to a larger area because of special arrangements when these states come into the Union. Subsequent court cases provided that, for historic reasons, the boundaries of Texas and Florida (along the Gulf of Mexico side) would be three marine leagues (about ten and one-half miles).[22]

Various states have their own rules, bidding and leasing systems, and management programs. At the present time, only Louisiana, Texas (in the Gulf of Mexico), Alaska and California have legislated offshore programs. This chapter will only focus on the federal "OCS" system.

The Outer Continental Shelf Lands Act gave the Federal Government jurisdiction over the leasing of mineral resources on the lands lying seaward of state waters. Because of its location, on the Continental Shelf, and outside the coastal state waters, this area was titled the outer Continental Shelf (OCS). The 1953 Act also established very general guidelines and directives for the managing of the resources of the OCS and the leasing of tracts for mineral exploitation—under the administrative responsibility of the Secretary of Interior.[23]

This Federal system is presently undergoing substantial change as a result of passage of the 1978 Amendments. Regulations are being promulgated to implement new provisions dealing with use of bidding systems, exploration, development, production, safety, oil spill and fishermen's damage liability and compensation. A new leasing program—describing the leasing schedule (time and location of sales) for the next five years is also being prepared.

The present OCS leasing system is generally administered by the US Department of the Interior. Regulations dealing with competition, bidding systems, and production rates are promulgated by the US Department of Energy. Regulations for safety in offshore operations and enforcement of safety rules are jointly administered by the Coast Guard within the Department of Transportation and the United States Geological Survey within the Department of Interior. Input and access by coastal states and state and local review of OCS activities are jointly supervised by the National Oceanic and Atmospheric Administration (NOAA) within the Department of Commerce and a special OCS Coordination office within the Department of Interior. Finally, oil spill liability and compensation is administered by the Coast Guard and fishermen's damage and compensation is handled by NOAA.

The OCS process begins with development of a leasing program and schedule. Tentative dates for sale of rights are listed and published and comments sought as to the relative concerns—expedited activity, environmental effects, competition, state and local government issues.[24]

The process then progresses to call for nominations of areas on the Outer Continental Shelf of the United States to be leased.

The Department of the Interior requests such calls from industry, coastal states (of the United States), units of local government, and the general public. The calls are designed to provide a basis for determining the area to be investigated for a proposed future lease sale.[25] The Department of Interior then selects a proposed lease area.

During this period, oil and gas companies seek authority to conduct pre-leasing drilling, to determine the possibility of hydrocarbon deposits. Until recently, such pre-lease drilling was authorized only for areas where the likelihood of deposits was low—off-structure—to gather only geological and stratigraphic information. Recently, the Department of Interior indicated it would allow on-structure pre-lease exploration.[26] Also during this period a two-fold environmental review is undertaken. First, studies are taken of areas to evaluate environmental sensitivity and to make certain predictions as to the potential effects of activity.[27] Second, the Department of the Interior prepares a draft environmental impact statement (EIS). This is because the holding of a lease sale constitutes a "major federal action significantly affecting the human environment" and requires the preparation of an EIS on such actions.[28] Preparation of the EIS and the environmental study are to be completed six months prior to a lease sale.

In the draft EIS, the possible environmental impacts of granting leases in the area under study are examined. Examining the impacts involves collecting basic data on the geology, climate, oceanography, biological environment and other natural phenomena of the area proposed for the sale. Data is collected on currents, tides, air and water quality, seasonal temperatures and winds, marine collections of plants and animals, wildlife on any land mass, the socio-economics of coastal land areas, commercial and sport fishing, shipping, navigation, military activities, and beach use in and around the area. An analysis is then made of the data collected to evaluate the probable effects gas and oil development in the area would have on these features of the environment. In particular, the risks of oil spills, weighed on a computerized projection of worst-case analyses, are taken into account.

Alternatives to a gas and oil lease sale in the entire area under consideration are also evaluated in terms of the environmental impacts they may have. These alternatives may include the deletion of certain tracts from the area, the inclusion of specific environmental and economic conditions on any or all needs, or cancellation of the sale.

Once the draft EIS is completed, it is submitted to the Environmental Protection Agency (EPA) and to the public for review and comment. A public hearing is held to give all interested parties a chance to express their views on the draft EIS and the proposed sale.

After comments have been received and the public hearing has been held, the Department of Interior refines the draft EIS into a final EIS. The final EIS takes into account information and views submitted to the Department of Interior in written comments or at the public hearing, as well as additional data which may have come to its attention. The final EIS is submitted to the President's Council on Environmental Quality and is made available to the public.

Taking into account the findings of the final EIS, the Department of the Interior makes a decision whether to hold the proposed lease sale. The resulting decision may be to hold the sale, cancel the sale, delay the sale, or modify the sale by deleting a number of tracts or by including specific environmental and economic conditions on any or all tracts.

Prior to an actual sale, the Department of Interior working with the Department of Energy determines what types of bidding systems will be used. Use of new bidding systems was one of the major reforms of the new OCS statute.[29]

Under the 1953 OCS Act, leases were generally to be awarded to the highest responsible qualified bidder, through competitive and sealed bidding procedures, on the basis of a cash bonus, with a fixed royalty of no less than twelve and one-half percent, or on the basis of a royalty, at no less than twelve and one-half percent, and a fixed bonus. In fact, with very few exceptions, bidding until now has been only on the basis of cash bonuses—whoever puts up the most "up-front" money. Many individuals questioned the wisdom of the traditional cash bonus bidding system, which placed a premium on large up-front capital expenditures, and which many urged tended to favor the large oil companies, giving them an unfair competitive advantage over the smaller independent oil companies.

The 1978 OCS Amendments provide for the use of new bidding systems, including variable royalty, variable net profit, fixed net profit, and work commitment bidding alternatives. It is hoped that these new systems will increase OCS competition, more efficiently allocate scarce capital, and increase ultimate resource recovery.

The use of these new bidding systems will require close cooperation between the Interior and Energy Departments, as the Secretary of the Interior retains the power to issue leases, while the Secretary of Energy has the new responsibility for promulgating regulations implementing alternative bidding systems in accordance with the DOE Act. Such regulations, including those regarding modifications of new bidding systems, are to be promulgated in advance.

Bidding systems other than bonus bidding, including royalty, net profit, work commitment, and nonenumerated systems are to be utilized in at least twenty percent and not more than sixty percent of the tracts offered for leasing in all OCS areas during each of the next five years. In utilizing the new bidding alternatives, a variety of considerations are to be taken into account.[30]

The Secretary of Interior is given the discretion to use[31] and to waive the minimum or maximum requirement for the use of bidding systems other than bonus bidding, based on determination that either the minimum or maximum percentage level is not consistent with the purposes and policies of the Act.

"To preserve and maintain free enterprise competition," the Attorney General is given the discretion to review the competitive effects of the sale and recommend appropriate action to the Secretary of Interior.[32]

In addition to establishing bidding terms, the Secretary of the Interior also publishes lease terms, including stipulations, that are his condition for accepting a bid. These terms cover technical issues such as rentals, environmental issues such as protection of resources and management issues such as suspension or review procedures.[33]

Once bidding is completed, the Secretary is free to accept the highest qualified bid or to reject all bids as insufficient. Acceptance of the bid completes the contract and a lease is awarded, with all the conditions, terms, and stipulations previously published.

After awarding of a lease, and prior to commencing exploration, the lessee must prepare an exploration plan.[34] This plan is to detail all proposed activities and is to be accompanied by a statement of expected impact on adjacent coastal areas. States, local governments and others have the opportunity to comment on the plan. The Secretary can approve the plan and thus allow drilling to commence or request modification in light of comments. He also may disapprove and plan and commence proceedings to suspend or cancel a lease.[35]

If a plan, as offered or modified, is approved, the lessee must secure necessary permits, and then conducts exploration either until a discovery or the end of his lease term.[36]

After a discovery, the lessee must prepare a development and production ("d&p") plan. Just as with an exploration plan, this is to be accompanied by a statement of anticipated impact onshore and is to be reviewed by all interested parties.[37] Two review procedures are possible before a plan can be approved. At least once in every previously undeveloped area, the decision whether to approve a development and production plan is to be considered a major federal action, and the Department of the Interior must again go through the process of preparing an EIS.[38] If an EIS is not required, formal comments must be submitted and then considered. The Secretary of the Interior may then, considering all information, approve the plan or order modifications of the plan, if he determines that the plan does not make adequate provisions for safeguarding the human, marine, or coastal environments. The Secretary can and must disapprove the plan if he determines that because of exceptional resource values in the marine or coastal environment, exceptional geological features, or other exceptional circumstances, the implementa tion of the plan would cause serious harm to the marine, coastal, or human environments, that such harm is not likely to decrease substantially within a reasonable time and that the harm outweighs the benefits of approving the plan. If the plan is disapproved on that basis, the lessee is given five years to submit an acceptable plan. If no plan is approved by the Secretary within five years, the Secretary must cancel the lease.[39]

Actual production can commence once a "d&p" plan is approved and necessary permits secured. Production continues as long as the lessee desires, assuming diligent and safe operations.

CONCLUSION

The procedures for federal management of OCS activities were substantially changed by the new Amendments. It is impossible to detail all the changes in this chapter. Congress and the Administration believe that the new provisions will lead to expeditious and safe operations.

NOTES

[1]P.L. 95-372, 92 Stat. 629 (1978); 43 U.S.C. 1801 *et. seq.*

[2]Estimates for undiscovered recoverable resources have been tabulated as follows:

OFFSHORE OIL	BILLIONS OF BARRELS	
	USGS	NPC
Gulf of Mexico	3–8	19
Pacific	2–5	17
Atlantic	0–6	6
Alaska	3–31	29
Total	8–50	71

OFFSHORE GAS	TRILLIONS OF CUBIC FEET	
	USGS	NPC
Gulf of Mexico	18–91	220
Pacific	2–6	25
Atlantic	0–22	60
Alaska	8–80	195
Total	28–199	490

This represents natural gas only. Undiscovered recoverable resources for natural gas liquids are estimated at 11-22 billion barrels for both onshore and offshore.

[3]States feared damaging impacts on their social and political infrastructure. Fishermen and tourist industries feared interference with their activities. Citizens and environmental groups feared massive oil spills would damage the marine and coastal areas. Some industrial and citizen groups believed that leasing resulted in little competition in offshore leasing. In general, complaints focused on the vagueness of the 1953 law; the inadequacy of present regulations; the failure to develop a comprehensive plan for leasing; the lack of compensation for spills and onshore impacts; the failure to consult with and consider the interests of all different levels of government, and interested groups and citizens; and the narrow, uncompetitive OCS lease bidding procedures. These fears led to several trends. First, states, local governments and citizen groups increasingly filed lawsuits. Second, as a result of suits and other protests, considerable slippage occurred in lease sale schedules. Third, the actual number of tracts actually offered for sale, the number actually bid on, and the number actually awarded were considerably below estimates.

[4]Statement by President Carter, at signing of Outer Continental Shelf Lands Act Amendments of 1978 into law on September 18, 1978. Administration of Jimmy Carter, Weekly Compilation of Presidential Documents, Vol. 14, at 1530-31 (Week Ending Friday, September 22, 1978). In his statement, the President "applauded the passage of (the 1978 OCS Amendments)" as "an important part of our energy program," he stressed that the new law "will provide the needed framework for moving forward once again with a balanced and well-coordinated leasing program to assure that OCS energy resources contribute even more to our Nation's domestic energy supplies."

[5]See J. Whitaker, *Striking A Balance—Environmental and Natural Resources Policy in the Nixon*-Ford Years at 259 (1975) [hereinafter cited as Whitaker].

[6]See A. Hollick, *U.S. Oceans Policy: The Truman Proclamations,* 17 VA J. INT'L. 23, 28-29 (1976).

[7]For an extensive discussion of these studies and the problems and policies involved in their preparation, review and implementation, see A. Hollick, *supra* note 6 at 29 to 55.

[8]*United States v California* 332 U.S. 19 (1947); *United States v Texas,* 339 U.S. 707 (1950); *United States v Louisiana,* 339 U.S. 699 (1950). As early as 1937, these states had begun to lease rights for oil and gas exploitation off their coasts. Whitaker at 260.

[9]Proclamation No. 2667, September 28, 1945, 59 Stat. 884. An Executive Order placed responsibility for leasing and management of these resources under the Secretary of the Interior. Executive Order 9633, 59 Stat. 885, 3 CFR 437 (1943-48 Compilation).

[10]Geneva Convention on the Continental Shelf, 3 U.N. Doc. A/Conf. 13/L 55 T.I.A.S. 5578.

[11]*United States v Texas,* 339 U.S. 707 (1950); *United States v Louisiana,* 339 U.S. 699 (1950).

[12]Whitaker at 261.

[13]The solicitor of the Department of Interior issued a formal opinion that the Mineral Leasing Act of 1920 did not apply to the Continental Shelf and, therefore, contained no authority to grant leases. Solicitor's Opinion, M-34985, 60 I.D. 26 (1954). See Whitaker at 260.

[14]P.L. 31, 83rd Congress, lst Session, 67 Stat 29, 43 USC 1301 *et. seq.*

[15]P.L. 212, 83rd Congress, lst Session, 67 Stat 462, 43 USC 1331 *35. seq.*

[16]Several statutes that peripherally affected OCS activities were enacted. The Fish and Wildlife Act of 1956, 16 U.S.C. 747 *et. seq.,* for example, provided that the United States Fish and Wildlife Service was to study, protect, and manage the fish resources under U.S. jursidiction.

[17]For a detailed discussion of the Santa Barbara blowout and its impact on Southern California, see *Outer Continental Shelf Oil and Gas Leasing off Southern California: Analysis of Issues,* prepared at the request of Hon. Warren G. Magnuson for the Senate Committee on Commerce, National Ocean Policy Study, 93rd Congress, 2nd Session (November 1974).

[18]See Organization For Economic Cooperation And Development, Environmental Directorate, Environmental Impacts From Offshore Exploration And Production Of Oil And Gas at 22-24 (1977) [hereinafter cited as OECD Report]; Whitaker at 266, n. 31.

[19]*Natural Resource Defense Council v Morton,* 458 F. 2d 827 (D.C. Cir. 1971).

[20]For a more detailed discussion of these accelerated leasing proposals, see Comptroller General, Outlook For Federal Goals To Accelerate Leasing Of Oil And Gas Resources On The Outer Continental Shelf (March 1975) [hereinafter cited as GAO - Accelerated Leasing Study]; *An Analysis of the Department of the Interior's Proposed Acceleration of Development of Oil and Gas on the Outer Continental Shelf,* prepared at the request of the Honorable Warren G. Magnuson, Natural Ocean Policy Study, 94th Congress, lst Session (March 5, 1975) [hereinafter cited as NOPS Accelerated Leasing Study].

[21]For a detailed description of the legislative history of the OCS Amendments, see the following reports:

(1) Senate Report No. 93-1140, Energy Supply Act of 1974, 93rd Congress, 2nd Session (September 9, 1974).
(2) Senate Report No. 74-284, Outer Continental Shelf Management Act of 1975, 94th Congress, 1st Session (July 17, 1975).
(3) House Report No. 94-1084, *Outer Continental Shelf Lands Act Amendments of 1976,* 94th Congress, 2nd Session.
(4) House Report No. 94-1632, 94th Congress, 2nd Session (Conference Report).
(5) Senate Report No. 95-284, Outer Continental Shelf Lands Act Amendments of 1977, 95th Congress, 1st Session.
(6) House Report No. 95-590, Outer Continental Shelf Lands Act Amendments of 1977, 95th Congress, 1st Session.
(7) Outer Continental Shelf Lands Act Amendments of 1978, House Report No. 95-1474 and Senate Report No. 95-1091 (August 10, 1978) (Conference Report).

[22]*United States v. Florida,* 363 U.S. 121 (1960) Louisiana, Mississippi and Alabama were not successful in their claim. *United States v. Louisiana,* 363 U.S. L (1960). Neither was California. *United States v. California,* 381 U.S. 139 (1960). With the extension of offshore leasing to the Atlantic, a renewed series of lawsuits commenced—this time by Eastern seaboard states. In 1975, the Supreme Court held that coastal state rights were limited to three miles for all states, except for the already granted exemption for Texas and Florida. *United States v Maine,* et al, 420 U.S. 515 (1975).

[23]In its administration of the OCS oil and gas program, the Department of Interior filled in details through its authority to promulgate rules and regulations. See generally 30 CFR 250.1 *et. seq.;* 43 CFR 2883.0 *et. seq.* and 3300 *et. seq.*

[24]Outer Continental Shelf Lands Act, as amended, §18 (hereinafter OCSLA). The Secretary is to consider all relevant opinions and values, prepare a proposal and submit it to Congress and others affected and then promulgate and follow a schedule. Specifically, the leasing program is to consist of a schedule of proposed lease sales indicating the size, timing, and location of leasing activity which he determines will best meet national energy needs. The program is to consider the economic, social and environmental values of the renewable and nonrenewable resources of the OCS and the potential impacts of OCS activities on the marine, coastal and human environments. In so doing, the Secretary shall attempt to strike a proper balance between the poten-

tial for environmental damages, the potential for the discovery of oil and gas, and the potential for adverse impact on the coastal zone. During the preparation of the proposed five-year leasing program, the Secretary is to invite and consider the suggestions from interested Federal agencies, including the Attorney General, in consultation with the RTC and the Governor of any affected State. The Secretary may also invite or consider any suggestions from the executives of any affected local government. Hence, the Secretary shall seek the input from a wide variety of sources and open up the OCS decision-making process regarding the formulation of a leasing program. In addition, the Secretary shall reply in writing to a Governor's request to modify the leasing program, and he may modify such program if he feels that it is appropriate. So that the Congress can monitor the development of the proposed leasing program, and insure that the recommendations of the Governor's and others are properly considered, all correspondence between the Secretary and the Governors of affected States and any other related information shall accompany the five-year leasing program when it is submitted to the Congress. Such program is to be submitted to the Congress, the Attorney General, and the Governors of affected States, and published in the Federal Register within nine months after the date of enactment. The Attorney General, after consultation with the RTC, is to submit comments on the effects of the leasing program upon competition. Furthermore, any State, local governments, or other persons may submit comments and recommendations regarding any aspect of the proposed leasing program. When the Secretary submits the proposed leasing programs to Congress, he shall indicate why any specific recommendation of the Attorney General or a State or local government was not accepted. No later than eighteen months after the date of enactment, no lease shall be issued unless it is for an area included in the approved leasing program.

[25]OCSLA, §18(f)(1).

[26]See 44 Federal Register 8302 (February 9, 1979). This is a highly controversial issue. Some companies argue that this is the "first step" to total government exploration and development – a Federal oil and gas company (FOGCO). Others urge that this type of pre-lease date – collected through exploration activity by private companies – is essential to compile and assess information on resources.

[27]OCSLA, §20.

[28]National Environmental Policy Act, §102, 42 U.S.C. 4332.

[29]OCSLA, §8(a).

[30]These considerations include, but are not limited to, (1) providing a fair return to the Federal Government; (2) increasing competition; (3) assuring competent and safe operations; (4) avoiding undue speculation; (5) avoiding unnecessary delays in exploration, development and production; (6) discovering and recovering oil and gas; (7) developing new oil and gas resources in an efficient and timely manner; and (8) limiting administrative burdens on government and industry.

[31]The OCS Act contains a provision providing that thirty days prior to a lease sale, a notice must be sent to Congress and published in the Federal Register, identifying bidding systems to be used in that sale, designating tracts to be leased under the various systems, and why a system is being used for a tract. Congress decided that particular attention had to be paid to the proposed use of bidding systems not enumerated in the Act. It may be necessary for the Secretary of Energy to employ modified or non-enumerated bidding systems. However, such systems can only be used if they are submitted to Congress and then only if neither House passes a resolution of disapproval within thirty days.

[32]Responsibility for the review of the competitive effects of a lease sale is given to the Attorney General and it is discretionary. However, the Attorney General is to consult with the Federal Trade Commission in making any review, in securing information and in making recommendations. The Attorney General, after consultation with the

FTC, may make recommendations as to whether a lease sale or sales indicate a situation "inconsistent with the antitrust laws." The Secretary of Interior is free to accept or reject any recommendations. He may accept a bid and award a lease despite any recommendation from the Attorney General, so long as he notifies the lessee and the Attorney General of the reasons for his decision. He may, of course, accept the recommendations and refuse to accept an otherwise qualified bid and not award a lease. There is no requirement of a hearing prior to this decision of this decision of the Secretary. In addition, the failure of the Secretary to accept a recommendation is not a basis for an independent cause of action against a sale or lease award.

[33]OCSLA, §8(b).

[34]OCSLA, §11(c).

[35]Cancellation is authorized by §5(a)(2) of the OCSLA. Prior to passage of the 1978 Amendments, cancellation for environmental reasons was not authorized. See *Gulf Oil Corporation v Morton,* 493 F.2d 191 (9th Cir. 1973); *Union Oil Company v Morton,* 512 F.2d 493 (9th Cir. 1975). Suspension and then cancellation is authorized at any time provided the requisite showing of a probability of serious harm or damage. Such an environmental cancellation entitles the lessee to compensation—the lesser of the value of the lease rights or costs.

[36]OCSLA, §5(b). The lease term is ordinarily five years but may be ten years in exceptional circumstances.

[37]OCSLA, §25.

[38]At least once, prior to authorizing production in any previously undeveloped area, an EIS must be prepared. OCSLA, §25(e).

[39]Again, cancellation is governed by procedures generally applicable to all OCS activities and compensation is to be paid. See discussion in note 35, *supra.*

What I have sketched comprises elements of an argument that leads one to question the linear extrapolations about change made by most observers and participants here. No matter whether one views this argument with alarm, as some do, or with the optimistic sense that such change represents a significant opportunity for man to redesign his political institutions to more closely approximate classic democratic theory ideals, is immaterial. If the argument has merit it puts discussion about marine policy questions in a new light. For example, discussion about marine policy within our postindustrial societies will be more—not less—complex in the future as economic and ecological tradeoffs are grappled with. However, it is in the comparative, international dimensions of marine policy that the implications of this alternative-development model of postindustrial societies and the collective goods language used to interpret the development pattern are of special interest.

Let us recall the description of the problem facing comparative marine planners in the Wenk chapter; that it is an accurate assessment there can be little doubt. The problem is that the nations not represented here, from the second and third worlds, are at points in the development process where they are willing to bear the costs of the negative externalities so well summarized by Wenk's chapter. They are willing—again in the language of economics—to internalize these externalities because their priorities remain material in orientation. Unless substantial incentives or sanctions are provided for these countries to cease their roles in destroying the oceans, eloquent statements of the Wenk variety will be regarded as liberal cant and ignored. Second, if the centralized political institutions of our postindustrial countries do continue to weaken, there will be declining leverage points to move the oceans from their present position as an example of the tragedy of the commons. And, finally, if we break down the goods provided by the oceans, some like pollution do affect countries differentially; pollution, until its advanced stages, is a collective, not public evil. Military problems, fishing, minerals, also affect different parts of the globe differently. Centralized planning units are unlikely to be the appropriate institutional design to handle the above problems. Rather, we should seek to work out more flexible institutional arrangements

that deal with functional areas of marine policy (see the Hennessey chapter).
In sum, at a time when calls for centralized, coordinated comparative marine policy are being made, the centralized national planning structures in postindustrial societies are in disarray. To move toward a centralized institutional design for the oceans would simply create another disjuncture between individuals and groups and the problems they must deal with.

CHAPTER 9

Offshore Oil and Gas Policy: United Kingdom

G. A. MACKAY

University of Aberdeen
United Kingdom

Wine that maketh glad the heart of man:
and oil to make him a cheerful countenance
(Psalm 104, v. 15)

INTRODUCTION

Given the international nature of this collection, I intend to concentrate on: (a) those aspects of policy which are markedly different in the United Kingdom (UK) compared with other Outer Continental Shelf (OCS) countries; and (b) those policies and experiences which may have relevance for OCS developments elsewhere, particularly the US. Thus I shall ignore aspects of purely domestic (or parochial) interest—for example, the location of onshore facilities such as landfall terminals and production platform sites—and other aspects of limited international significance—such as health and safety legislation, trade union legislation, and the like. This is not to say that these are unimportant in the UK; I would be happy to provide interested parties with more detail on these elsewhere.

There are five aspects on which I would like to concentrate: (1) exploration and licensing policy; (2) production and depletion policy; (3) taxation policy; (4) the state-owned (public) British National Oil Corporation (BNOC); and (5) the conflict between offshore oil and fishing.

First, it is necessary to provide some brief background information. OCS development in the UK dates back only to 1965 and offshore oil production first began in June 1975, so we have a very brief history in comparison with the US. Nevertheless, the North Sea is the deepest and harshest offshore area in which exploration and production have taken place, so there is a good deal to be learned from Norwegian and UK experience, particularly in offshore and underwater technology. At present we are producing from fields up to one hundred fifty miles offshore and in water depths up to six hundred feet.

The pace of development since 1965 has been very fast. Until 1975 all our oil requirements were imported, mainly from the Middle East; there was no domestic natural gas, although some gas was produced from coal; coal was the main energy source. Oil now provides about fifty-five percent of our energy requirements, and offshore oil more than half of that; another ten percent comes from offshore gas production; our oil imports have been halved in quantity terms, although thanks to OPEC* not in monetary terms; and by 1980 we should be self-sufficient in energy production. Thus in five years we shall have moved from a substantial importer of oil to a net exporter, with consequent transformations in our balance of payments and government revenue. The table below sets out the offshore oil production since 1975.

Oil (million tons)

1975	1976	1977	1978	1979 est.	1980 est.
1.1	11.6	37.3	53.5	80.0	100.0

In comparison, domestic oil consumption in 1978 was eighty-five million tons.

The final introductory point I would like to make concerns the political system. Since the Second World War there have been only two political parties in government—the Conservative and Labour parties, which can in simple terms be described as pro-business and pro-labor respectively. The Labour (Socialist) Party was in power from 1964–1970 and again from 1974–1979; the Conservative Party was in power from 1970–1974 and was re-elected in May 1979. Both parties have had markedly different attitudes and policies towards OCS development and hence there have been frequent changes in policy. Indeed, some of this paper could be soon outdated when the new government has had time to reformulate its policies. Unfortunately, we have not been in the position of the Norwegians in having a fairly stable government and general agreement among the various political parties.

EXPLORATION AND LICENSING POLICY

The UK sector of the North Sea is divided for licensing purposes into blocks of approximately one hundred square miles in size. UK waters outside the North Sea have been similarly sub-divided, although final agreement has not been reached on other international boundaries. The

*Organization of Petroleum Exporting Countries

first licenses for oil and gas exploration and production were issued in 1964 and to date there have been six license rounds: in 1964 (348 blocks); 1965 (127); 1970 (106); 1971–1972 (282); 1976–1977 (44); and 1979 (42 blocks). Thus there have been two lengthy periods when no new licenses were issued: 1965–1970, which was a period of limited interest in the North Sea because only small gas discoveries were made, (the first major oil discoveries being the Ekofisk field in Norwegian waters in 1969 and the Forties field in UK waters in 1970, both of which generated a great deal of new interest); and the period 1972–1976 following the very large fourth license round.

The policy considerations which dictated the first round of licensing, and which have been subject to only minor qualification subsequently, were as follows:

> First, the need to encourage the most rapid and thorough exploration and economical exploration of petroleum resources on the Continental Shelf. Second, the requirement that the applicant for a licence shall be incorporated in the United Kingdom and the profits of the operation shall be taxable here. Thirdly, in cases where the applicant is a foreign-owned concern, how far British oil companies receive equitable treatment in that country. Fourthly, we shall look at the programme of work of the applicant and also at the ability and resources to implement it. Fifthly, we shall look at the contribution the applicant has already made and is making towards the development of resources of our Continental Shelf and the development of our fuel economy generally.[1]

Quick exploitation was the main driving force of British licensing policy, and from this many other consequences flow. The perceived need for rapid development determined the method of allocating licenses, the reliance on foreign capital and the financial terms initially established. The successful applicants were not determined by a competitive auction where the license went to the highest bidder, but by administrative discretion based on the proposed work programs of the applicant. Other things being equal, the applicant with the most intensive work program was successful. Rapid exploration was further encouraged by requiring each licensee to surrender one-half of his area after a six-year period, so that there was an obvious incentive to determine which part of any area offered the best commercial prospects. Finally, the annual rental of each block licensed was low. In the first round it was fixed at £6250 per annum for the first six years, rising subsequently in £10,000 stages to reach a maximum of £72,500 per year, and although the rent was raised subsequently, in the third and fourth licensing rounds, it remained almost nominal. In this way, only a minimum strain is put on the working capital of the oil companies and this enables them to finance rapid exploration.

Speed dictated a reliance on foreign-controlled oil companies, particularly American companies. It was considered that the two UK, or partly UK, majors—British Petroleum and Shell—did not have the necessary resources to develop all the licensed areas as quickly as was desired, but some preference was shown to UK companies, as they appear to have obtained a relatively higher proportion of the more promising areas.

In the second licensing round (1965) the new Labour administration added two further criteria. First, it would consider the contribution which applicants had made to the UK balance of payments and to creating employment in the United Kingdom, with particular reference to regional considerations and, secondly, "proposals which may be made for facilitating participation by public enterprise in the development and exploitation of the resources of the Continental Shelf"[2] would be taken into account.

In the first three rounds all licenses were allocated by the administrative discretion of the Department of Trade and Industry (now the Department of Energy) and the fifth and six rounds have been likewise. Competitive licensing was not attempted until the fourth round, and even then was restricted to only fifteen of the 286 blocks eventually allocated. The arguments for discretionary licensing can be quickly summarised. The UK was heavily dependent on external sources of fuel, and it was desirable to provide a secure and—hopefully—cheap source of fuel from indigenous sources, which would benefit the balance of payments. The North Sea offered such a possibility, but it was an unproved area which could be tested only by expensive exploration and drilling carried through at the limits of existing off-shore technology. The licensing system adopted allowed the Department of Trade and Industry to exercise its discretion in favor of companies that would pursue an active search program. It had the further alleged advantage that it allowed some (concealed) discrimination in favor of UK operators, and overt pressure could be placed on all operators—UK or foreign—to buy British goods and services for off-shore exploration and production.

There is little doubt that the major objective was met. Exploration has been more rapid and more extensive in the UK sector of the North Sea than in any other national sector, although a major contributory factor has been the high success rate achieved in UK waters.

Given Britain's dependence on external fuel supplies, a dependence underlined by the rise in the price of Middle East oil, it is clear that speedy exploitation was extremely important. This the discretionary system did achieve. Speed was almost certainly at the price of some short-term loss to the Exchequer, as a competitive system of licensing would almost certainly have resulted in a higher annual rental per block licensed. However, given the very imperfect knowledge of the geological conditions prevailing and the limited experience of operating in North Sea conditions, the potential loss of immediate revenue could not have been absolutely large, and was relatively extremely small compared with the gains of quick exploitation.

While this appears to be the correct assessment for the first three licensing rounds, it is difficult to resist the conclusion that the continued heavy reliance on the discretionary system in the fourth round was mistaken. It could be defended on the grounds that the increase in the posted prices, levied by the OPEC countries in September 1970, further underlined the need for haste; and there was also concern at what appeared to be a fall-off in exploration activity from 1970—in consequence, it would appear, of a rising dry-hole ratio. On the other hand, the major Ekofisk oil strike had been announced in November 1969. In the UK sector the discovery of the Forties field was announced in 1970, the more minor discoveries of what later turned out to be a major field had been drilled before the fourth round. Certainly, when the fourth licensing round was completed, the results indicated that the oil companies and the Department of Trade and Industry (DTI) had very different assessments of likely prospects. The 271 blocks allocated through the discretionary system brought an income of only some £3m. The handful of fifteen blocks, which had been deliberately selected by the DTI as a "representative cross-section" of the whole area to be licensed, was sold by competitive auction for £37m.[3] In short, the DTI took a conservative, even a pessimistic view of the future, while the oil companies were evidently more optimistic. Subsequent events have proved the assessment of the oil companies to have been correct.

A great deal has been written on the relative merits of competitive bidding and administrative licensing, for example, by Kenneth Dam.[4] My own view—explained in more detail elsewhere—is that both sides of the argument have been exaggerated, but I have a slight preference for the auction system. Nevertheless it is unlikely that this aspect of UK policy will change in the foreseeable future.

PRODUCTION AND DEPLETION POLICY

During the first ten years of activity on the UK Continental Shelf the policy of successive governments was to exploit reserves of oil and gas as quickly as possible. Companies with successful discoveries had a free hand in developing fields; there were no significant controls over production levels.

When the Labour government returned to power in 1974 it became clear that there were second thoughts about the appropriateness of a rapid depletion policy. In part this was due to pressure in Scotland from the Scottish National Party, who wanted policies more in keeping with Scottish needs than UK needs. A government white paper (consultative document) in 1974 announced that the government intended to take powers to control the level of production in the national interest, not with a view to the immediate period ahead but with a view to the 1980s. This attracted a lot of criticism from oil companies, and the Secretary of

State for Energy issued a series of guidelines which were subsequently incorporated into the 1975 Petroleum and Submarine Pipelines Act. The guidelines are as follows:

1. For finds made before end-1975 under existing licenses: No delays would be imposed on development plans. Any production controls would not be applied before 1982, or four years after the start of production, which ever is later.
2. Finds made after end-1975 under existing licenses: No production cuts will be made before 150 percent of investment in the field had been recovered.
3. All finds made after end-1975: Development delays would only be imposed after full consultation with the companies concerned, so that premature investment is avoided.
4. General: Any use by the government of depletion controls would recognize technical and commercial aspects of the fields in question which would generally mean production cuts of no more than 20 percent. The industry would be consulted on the necessary period of notice before cuts became effective. The government would also take into account the needs of the offshore supply industry in considering development delays or production cuts.

So far these powers have only been exercised in relation to oil fields with significant reserves of associated gas. Companies have been told to delay or reduce oil production until such time as they have agreed to plans with the Department of Energy for the satisfactory handling of the gas. This was done to avoid excessive flaring of associated gas, and the agreements reached appear very sensible. Flaring is still allowed in some cases but in others the proximity of the Frigg and Brent gas pipelines has enabled plans to be formulated for the use of the gas.

More informal pressure has been put on a few companies in relation to their oil development programs. Mesa, the operator for the Beatrice field (a small field in the Moray Firth only fifteen miles from shore and in a good fishing area) was refused permission for offshore loading and had to resubmit poposals, which included a pipeline to shore. Permission has been similarly delayed for the Maureen and Magnus fields, but in no case does it appear that an excessive level of production—either on an individual field or aggregate basis—was a factor.

With the new Conservative government, it is almost certain that these controls will be scrapped and that the oil companies will again be given a free hand in developing their fields. Even if oil production reached one hundred fifty tons per year by the mid-1980s—which is probably the maximum possible—this would only amount to a fifty percent surplus over domestic needs, and the UK balance of payments is such (and almost certainly will still be such in the 1980s) that the additional foreign revenue will be a major benefit to the national economy. There is also the pressure to maintain a steady level of demand for production platforms, modules, pipelines and other equipment.

This means that control over production could only come indirectly through the regulation of exploration activity, which is the main method of control in Norway. In fact, comparisons are often made in the UK with the Norwegian's "go slow" policy, but these are misleading. In relation to known reserves, there is little difference between UK and Norwegian policies. At present planned production levels, proven reserves of oil and gas in both sectors will be finished at about the same time—twenty to twenty-five years hence. However, in relation to domestic requirements, the UK target level of one hundred fifty million tons is about fifty percent greater than domestic demand. In Norway the target level of ninety million tons is about eight hundred percent greater. Looked at in this way, it could be argued that the Norwegian rate is about forty-four times faster than in the UK.[6]

The Labour government's depletion proposals attracted much criticism in the UK for two main reasons—the uncertainty they created, and their economic rationale. Regarding the latter, there is no need here to provide a detailed account of the theory of resource depletion. A quote from two British economists, Robinson and Morgan, will suffice:

> It is an open question whether, at any given time, government can even identify in which direction company programs should be varied, let alone whether it can determine such an elusive question as the optimal depletion rate for society, the criteria for establishing which are by no means obvious.... One must consider the possibility that government intervention will move the depletion rate further from its optimum than it would have under producer control.[7]

So far I have concentrated very heavily on oil, and ignored gas production, which is much less important in the UK. Nevertheless it is necessary to discuss it briefly in the present context, because policies for oil and gas differ considerably. Offshore gas production in fact began in 1968, and production at present averages about thirty-five million tons oil equivalent per year. This will probably be the average level throughout the 1980s, after which production is likely to fall sharply. In comparison, offshore oil production in 1978 was 53.5 million tons, and in the 1980s should range 120–150 million tons per year.

All natural gas produced in the UK (both offshore and onshore) has by law to be sold to the state-owned British Gas Corporation (BGC) which to all intents and purposes is both a monopoly purchaser and seller. We have an extensive natural gas grid for domestic and industrial consumers, and this has been expanded very rapidly since the discovery of gas in the North Sea. Because of its monopoly position, BGC has been able to offer companies a low price for gas, with the exception of gas from the Norwegian part of the joint UK-Norwegian Frigg gas field. This has generated a great deal of argument.[8]

In brief, the UK is now in a very good position regarding gas, because potential supplies are in excess of demand (which is being held down by the government to encourage coal consumption and production). The main policy of the BGC at present is to take associated gas from oil fields coming into production (for example, the Brent field) and to reduce production from existing non-associated fields (for example, in the southern North Sea) and to delay production from new non-associated fields. Again this has attracted a great deal of criticism, but is probably sensible given the greater importance of oil to the UK economy.

Finally, one related aspect is the control over the destination of any oil produced. By law, any offshore oil has to be landed in the UK, unless special permission is given by the government. Landing does not mean refining or processing, however, and at present about forty-five percent of production is exported in crude form, mainly to the US. The two main landfall terminals-Flotta in the Orkney Islands and Sullom Voe in the Shetland Islands-are really just trans-shipment and storage terminals.

In the aftermath of the Iranian troubles and the consequent supply problems in the Middle East, there is increasing pressure in the UK for more of the oil to remain in the UK, and it is possible that some controls over refining and marketing could be introduced, probably just as a temporary measure.

TAXATION POLICY

The main OCS taxes are – as follows: (1) Royalties at 12½ percent of the wellhead value; (2) Corporation tax at fifty-two percent of taxable profits (this is the normal tax paid by UK industry; and (3) A special Petroleum Revenue Tax (PRT) introduced in 1975 at forty-five percent of taxable profits.[9]

Regarding PRT, there is a capital "uplift" allowance of seventy-five percent (in other words 175 percent of capital expenditure in a given period) before profits are assessed; a ceiling on the rate of tax; a safeguard provision relieving fields of tax if the rate of return falls below a defined level; a half-year allowance equivalent to 500,000 tons (1,000 barrels per day) before the tax becomes due; and a discretionary ability of the government to reduce royalties.

PRT was introduced to try to eliminate the windfall profit accruing to oil companies as a result of the OPEC price rises in 1973–1974 also 1978–1979. Its effect is that the total government tax take is around seventy percent although it does vary considerably from field to field. The early fields—such as Forties and Piper—are certainly very profitable, because much of the capital expenditure was incurred before 1975; there are signs that some of the newer and smaller fields are becoming marginal.

UK government tax revenues in 1978 amounted to £442 million

($900 million). By 1985 they should increase to around £5,500 million per year ($11,000 million) (in 1979 prices), with the composition being Corporation Tax forty-five percent, PRT and royalties twenty percent. Again the government take of around seventy percent has attracted criticism from the industry, but most independent observers believe that it is a reasonably fair system. The main independent criticisms are that the system is rather inflexible and does not encourage the development of small, marginal fields (although in that respect it is not as bad as the Norwegian system).[10]

THE BRITISH NATIONAL OIL CORPORATION (BNOC)

Another innovation by the Labour government in 1976 was the establishment of a state-owned BNOC. In part it was modelled on the Norwegian equivalent, Statoil, which was set up in 1972. BNOC started operations in January 1976 so its three-year life is considerably shorter than that of Statoil, although already it is much bigger. This rapid rate of growth, when compared with Statoil, is largely attributable to the fact that in 1976 BNOC took over most of the existing North Sea operations of Burmah (the British oil company which was rescued from severe financial difficulties – probably liquidation – by the government) and the National Coal Board, and thus inherited stakes in a number of fields as well as Burmah's qualified and experienced staff. At present BNOC's staff number about one thousand and is expected to increase to over two thousand by the early 1980s; Statoil currently employs about six hundred.

As with Statoil the remit given to the UK state oil company is surprisingly vague. It was set up as part of the series of legislative changes made in the Petroleum and Submarine Pipelines Acts (1975), which included major changes in taxation and depletion policies. As set out in its first annual report[11] BNOC's main priorities are as follows:

1. The efficient and commercial management of its equity interests in exploration, development and production.
2. The effective disposition of petroleum available to the Corporation, both from its equity interests and through participation arrangements with other offshore petroleum production licensees, with due regard both to commercial considerations and to national and international interests and obligations.
3. The development of expertise in and knowledge of all aspects of the oil business, but particularly those relating to the development and use of resources under the UK continental shelf.
4. The development of its capability to give informed and expert advice on oil matters to the Secretary of State as a contribution to the development of national policy.

One of Britain's leading financial journalists, Adrian Hamilton, has described the situation thus:

> The creation of BNOC was viewed with some suspicion by both the Department of Energy, because it might diminish its powers of control, and the Treasury, which was worried about cost. It raised fears in the industry that it could distort prices and returns by competing in the market place on a non-commercial basis, and it aroused the traditional parliamentary conflicts over state intervention. The solution, as in the other cases, was found in a compromise form of unusual qualifications and an unusual lack of definition... The question of refining and marketing was left open with a vague statement that it would not occur for some years anyway.[12]

Under Part 1 of the 1975 Act BNOC is charged with carrying out the normal functions of an integrated oil company, including exploration, production, refining, distribution and petrochemical production throughout the world. In addition it has the duty to advise and inform the Secretary of State for Energy on oil matters and "to perform services on behalf of the Crown." Policies and operations are decided by the Corporation's members (equivalent to a private company's board of directors but appointed by the government), two of whom are civil servants, and the Secretary of State has the power to direct the activities of the Corporation.

In 1976 BNOC inherited the National Board's offshore oil and gas interests, which included gas production from the Viking field and equity interests in four oil fields under development—Thistle, Dunlin, and the UK parts of Murchison and Statfjord. Through Burmah it acquired a share in the Ninian field and an additional stake in Thistle, and in fact became the operation for the Thistle field. Participation agreements were also reached with a few companies and over the last three years these have been extended to include all companies in the UK sector. For licenses issued during the first four UK rounds, all these agreements have been voluntary and on the basis of a "no gain, no loss" principle from the point of view of the companies involved. Originally it had been intended that this would mean BNOC contributing its full share of past costs, but some companies were unwilling to accept this; the normal procedure now is not for BNOC to have an equity stake but to have guaranteed access to fifty-one percent of production, to be purchased at market prices if BNOC wishes to do so. In the fifth license round this condition was included automatically and in the sixth round offer documents (applications for which were made in November 1978) potential licensees were invited to offer BNOC more than fifty-one percent share and to carry all or part of the Corporation's exploration.

In addition to these equity and participation interests, BNOC was given exclusive licenses for some blocks, although recently it has been obliged to offer Burmah shares in some of these. Its first discovery as an operator was made in June 1978 on block 30/17 b; this may prove to be a small field with around two hundred million barrels recoverable reserves.

At present BNOC is disposing of around 150,000 barrels per day (bpd) and this should increase to around one million bpd by 1981 – probably over forty percent of UK oil production by then. Of this approximately fifteen percent will come from BNOC equity shares, thirty percent as royalty oil and fifty-five percent from participation agreements. This is not necessarily a profitable aspect of operations, particularly if prices are static or falling – but it reflects a major concern of the UK government to have control over oil supplies, in case restrictions on Middle East supplies ever reappear. Thus although a large part of British production is exported at present, BNOC could ensure that it was diverted to the domestic market if the need arose. This is in marked contrast to Norway where, of course, domestic consumption is very small and security of supply not a major issue.

Another important difference between Statoil and BNOC is the former's greater involvement in downstream activities. Again, this largely reflects differences in economic structures: the UK has a substantial domestic refining industry in which British companies such as BP and Shell have major shares and at the present there is no obvious gap which BNOC could usefully fill.

There are also contrasts in information and regulatory aspects, although in both countries the full significance of these will not be clear for some years. The UK government has always been able to obtain information and advice from BP, and Shell to a less extent, so BNOC has a lesser role to play than Statoil. This is probably true of regulation as well, although the 1975 Act suggests otherwise. BNOC itself has spoken out strongly against its having regulatory powers – because of the danger of conflicts of interests – and maintains that this is the proper role of the Department of Energy.

An example of this conflict has already arisen with the 30/17 b discovery mentioned above. It has been claimed that this structure had been identified by Shell-Esso from seismic surveys in late 1975 and that the block had been their top priority in applications for fifth round licenses, but that BNOC had been given the major share in the block. The implication was made that BNOC only "discovered" the field because it had had access to the seismic data which Shell-Esso were obliged to pass on to the Department of Energy – data which they regarded as confidential. A letter written under a nom de plume appeared in *The Times* claiming that "One cannot give much credit to a person having the privilege of viewing all the hands in a poker game, then declaring himself a partner of who ever holds the best hand and finally, after settling down to play the hand, telling everyone what a good player he is as he rakes in the chips."

This particular allegation was refuted but it led to the Minister of State for Energy stating publicly that the Corporation had to recognize the importance of "compartmentalising" its operations—of keeping its monitoring of confidential information separate from its commercial operations. This conflict—or potential conflict—has been worrying oil companies who are concerned about BNOC using information gained through participation agreements to enhance its own commercial operations. In some cases special clauses have been written in to state participation agreements in order to safeguard companies against the possible misuse of information.

Another example of conflict arose in discussions about the use of the Ninian pipeline through which the group developing the Heather field wished to pass their oil. BNOC was involved in two ways—with an equity interest in the Ninian field and an option to participate in Heather. In this instance BNOC withdrew from the negotiations.

It is too early yet to assess the performance of BNOC, but the political attitudes of the new Conservative government are such that its powers are likely to be curtailed. Many people in the government believe that BNOC is and was unnecessary because of the government's majority shareholding in BP. It is also fair to say that many of the original supporters of BNOC have become disillusioned because it has acted in a manner little different from any private oil company and appears to have few "socialist" intentions. Statoil, of course, is in a very different position because Norway had no major national company like BP or Shell.

THE CONFLICT BETWEEN OIL AND FISHING

The final aspect with which I would like to deal concerns the fishing industry. This may appear rather peripheral in relation to the preceding discussion, but it is an issue on which I have been working recently and it fits in well with the overall motivation of this conference.

In the work I have been doing in Aberdeen on the oil versus fish conflict, I have been impressed by attitudes and policies in the US. This may surprise the fishing interests here; but it is probably more a reflection of the appalling lack of consideration given in the UK to the fishing industry.

There are four main aspects of conflict: loss of access to fishing grounds; damage to nets and gear from oil-related debris on the sea bed; pollution from spills; and the movement of labor and vessels away from the industry. This is not the place to discuss these in detail, but they are all important. Luckily oil pollution in the North Sea has not been a serious problem to date, except in a few localised incidents such as the spill at the Sullom Voe terminal. There is a debris compensation fund but to date total payments have been less than £80,000 ($160,000), a very small sum. On loss of access, which is a serious problem in the oil provinces east of Shetland and in the Moray Firth, there has been no agreement and fishing interests have been completely ignored.

Sadly, this lack of cooperation stems mainly from government officials (including those in the fisheries departments!) rather than the oil industry. The government officials are keen to maintain the principles of "free access" and "public rights" despite the fact that some of the offshore oil legislation contradicts those principles.

The problems have reached the stage that the fishing industry is on the point of taking protective action, which I think is a pity because I am sure that without government interference the oil and fishing industries would reach an amicable agreement. This action could take the form of legal action preventing the allocation of licences in important fishing areas—which I understand has already happened in the US. It is certainly one aspect of policy from which US experience can teach us.

CONCLUSIONS AND KEY ISSUES

I hope that the preceding points have highlighted the more interesting aspects of UK policy. Given the basic objectives of policy, I think our experience has been successful. Most of the initial objectives have been achieved. There have been problems, of course, but these have been minor on the whole; and there have been agreements, for example between the government and the oil companies, but they are understandable.

Looking to the future, I think there are five key issues:

1. How to encourage continuing exploration in the face of rapidly rising costs and declining success rates.
2. How to maintain a fairly stable level of offshore activity in order to avoid the problems of sharp fluctuations in the demand for production platforms, equipment, labor, and the like.
3. How to maintain some security over indigenous oil and gas supplies in the face of recurring problems in the Middle East. In this context the future role of BNOC is important.
4. How to ensure a fair balance in the share of financial benefits between the government and private oil companies.
5. The resolution of the conflict between offshore oil and fishing interests.

The honeymoon period in the UK is over. We are moving from a period of hectic youth to a more stable middle age, and there is no guarantee that policies designed for the former are the best for the latter. Issues one and three above suggest that some changes in policy are required in the near future. It is to be hoped that the turnover of governments in the UK does not prevent a stable and rational set of policies from emerging.

NOTES

[1]Minister of Power. *House of Commons Debate,* March 1964.

[2]Minister of Power. *House of Commons Debate,* July 1965.

[3]Public Accounts Committee. *North Sea Oil and Gas,* 1973.

[4]K.W. Dam. *Oil Resources: Who Gets What How?,* 1976.

[5]D.I. Mackay and G.A. Mackay. *The Political Economy of North Sea Oil,* 1975.

[6]T. Lind and G.A. Mackay. *Norwegian Oil Policies,* 1979.

[7]C. Robinson and J. Morgan. *North Sea Oil in the Future,* p. 33.

[8]For more detail see A. Hamilton *North Sea Impact,* 1978.

[9]For more detail see A.G. Kemp. *The Taxation of North Sea Oil,* 1976.

[10]A.G. Kemp and D. Crichton. *Taxation of Oil in Norway,* 1978.

[11]British National Oil Corporation. *Report and Accounts,* 1977.

[12]A. Hamilton. *op, cit.,* p. 39.

CHAPTER 10

Oil and Gas in Norway

EINAR RISA
Statoil
Norway

HISTORY

The first oil company approached the Norwegian Government in the fall of 1962, requesting permission to start petroleum exploration on the Norwegian Continental Shelf. The reason why this area was considered a possible oil and gas province was the discovery of gas in Groningen in the Netherlands toward the end of the 1950s. As a result of this request, Norway in May of 1963 proclaimed sovereignty over the Norwegian Continental Shelf with regard to exploration for and exploitation of submarine natural resources. The international legal basis for this decision was the Fourth Geneva Convention of April 1958, which gives the coastal states rights over the seabed and subsoils as far as such exploration and exploitation are concerned. This sovereignty extends as far as the resources can be exploited commercially, but not beyond the medium line relative to other countries.

According to this definition the southern part of the Norwegian Continental Shelf has a common border with the continental shelves of other North Sea countries like the UK, Denmark and Sweden. Its northern part extends as far as it is technologically possible to exploit it, except for the border area with the USSR in the Barents Sea. At the present time we are engaged in negotiations with the Soviet Union to decide on the dividing line between our two continental shelves.

The first reconnaissance licenses, which do not include the right to drill, were issued in 1963, while the first production licenses were awarded in 1965. The first exploratory drilling started in July 1966. The first commercial find was made in December 1969, and commercial production started in July 1971.

As of the end of 1978 drilling had started in 156 exploration and fifty-two appraisal wells. Of these 208 wells, 117 are located at a water depth of less than one hundred meters, 87 between one hundred and two hundred meters, and four at a depth of more than two hundred meters. The deepest well, in terms of water depth, has so far been drilled at a depth of 304 meters, that is, close to one thousand feet. When we look at the activity level on the Norwegian Continental Shelf, we have to consider the cost of operations in these climatically rather difficult areas. The average cost per well drilled in 1978 was more than five million dollars, which is considerably more than you have been accustomed to on the US Continental Shelf.

These exploratory activities have resulted in proven recoverable reserves at today's prices and available technology, amounting to 810 million tons (six billion barrels) of oil, thirty million tons (225 million barrels) of natural gas liquid and 26.5 trillion cubic feet of natural gas. About seventy-five percent of the oil and a little more than half of the natural gas have been developed or are under development. The question of developing the rest of these resources is under consideration. An important question in connection with these additional reserves is finding an economically acceptable transportation system. Because of the high development costs in the North Sea, an oil or gas field has to be of considerable size to be commercially exploitable.

The Norwegian petroleum production, of which all takes place on the Continental Shelf, in 1978 reached 30.6 million tons of oil equivalents (600,000 barrels per day) (oil and natural gas). This year the production is expected to reach about forty million tons (800,000 barrels per day), and about sixty-five million tons (1.3 million barrels per day) from 1981 until the middle of the 1980s, after which it will start to decline, unless new production is brought on stream. But we hope that the exploration which will take place during the next couple of years will provide the basis for a continuation of this level of production.

When we consider the prospects for new commercially exploitable petroleum finds, we have to take into account that the commerciability of a given field is dependent on a number of factors, the most important of which are as follows: (1) the size of field; (2) the recovery rate; (3) the geological structure; (4) the technology available for its development; (5) the transportation system available; (6) the price of oil and gas; (7) the markets available.

Some of these factors change over time, a few rather rapidly—like the last few months have shown us—when it comes to the price of oil.

THE INSTITUTIONAL/LEGAL STRUCTURE

The legal basis for the organization of the Norwegian petroleum activities is the "Act of 1963 relating to the Exploration for and Exploitation of sub-marine natural resources." It allows the Norwegian State the rights to sub-marine natural resources. This act also forms the basis for a set of regulations covering the various aspects of the petroleum activities on the Norwegian Continental Shelf.

The main decisions regarding our petroleum policies are made by the Parliament. The responsible government ministry in most petroleum questions is the Ministry of Petroleum and Energy. Safety questions, though, are handled by the Ministry of Municipal and Labor Affairs, and most petroleum taxation questions by the Ministry of Finance. The Ministry of the Environment is also involved in a number of questions. The Petroleum Directorate is the government's technical arm in these matters. It has wide inspection authorities and is charged with keeping the Ministry informed about the activities on the shelf.

Statoil—the Norwegian State Oil Company, Inc.—is a joint-stock company, organized in accordance with the Companies Act. All the shares are owned by the government, and the Minister of Petroleum and Energy is the company's general meeting. The corporate purpose of the company is to carry out exploration, exploitation, transportation, refining and marketing of petroleum and related products, and other closely related activities. This can be done alone or in cooperation with other companies. Statoil is thus an integrated oil company, charged with taking care of the state's business interests in the petroleum sector.

THE LICENSING SYSTEM

I think it might be worthwhile to go somewhat into detail about our licensing system, since this is one of the main instruments of the Norwegian petroleum policy, and since it is quite different from the system you use here in the U.S.

We use three main kinds of licenses:

Reconnaissance licenses give the licensees the right to do general geological and geophysical surveys, but not drilling. Such a license is non-exclusive, and does not imply any preference as far as production licenses are concerned.

For the purpose of awarding *production licenses* the Norwegian Continental Shelf south of 62N° has been divided into blocks of in most cases about five hundred square kilometers (193 square meters.) So far oil exploration has been restricted to these southern waters. Next year, however, drilling will start in a couple of areas north of this line. Production licenses have by now been granted in four rounds. The number of licenses granted has varied greatly from round to round. In the beginning each license consisted of several blocks. Now a license normally consists of just one block. The applicants—single or groups of companies—apply for one or more blocks after a notice for applications has been published. Then negotiations take place between the companies in question and the Ministry of Petroleum and Energy. Based on these negotiations, licenses are awarded to groups of companies, including one operating company, and two or more others. In each license Statoil, as the company taking care of the state's business interests in this area, holds a minimum of fifty percent. The choice of companies to be included in a license is made on the basis of many factors, like geological competence with respect to the expected problems to be solved in connection with that particular block, past performance, willingness to undertake a certain work program on the block, and willingness to be engaged in other industrial projects which could contribute to the development of the technological expertise of Norwegian industry. The work program I just mentioned includes drilling a certain number of wells within a defined timespan. It is rather obvious that the conditions which the government is able to impose on the oil companies involved, have something to do with the prospects for oil and gas in the area. Since finds were made in the North Sea, the international oil industry has become more eager to get licenses on the Norwegian Continental Shelf, and thereby willing to accept stricter conditions than was the case in the first phase of this venture. Furthermore, the Norwegian petroleum administration has built up considerable experience and technological expertise during the years these activities have taken place.

If Statoil or another Norwegian company is appointed operator for a license, one of the experienced foreign participants in that license becomes technical assistant during the exploration phase. This is to ensure that the best foreign technology is made available to the operator. To minimize economic risk for the Norwegian state in connection with the rather costly exploration for hydrocarbons, the other companies involved normally pay Statoil's share of the exploration costs. If a commercial find is made, however, Statoil pays for its share of the development costs. And I think it should be pointed out that the exploration costs, although high, constitute only a small portion of the total development costs for a petroleum field in the North Sea—about one percent

The third type of licenses concerns *building* of transportation systems, storage facilities and other equipment needed to develop a petroleum field. Such licenses have to be granted in each separate case when a field has been declared commercial. These licenses are to ensure that a field is developed in the most rational manner from the point of view of Norwegian society. For instance can the Government decide that a given pipeline is to be used also by others than the owner? This is to ensure an optimal utilization of the infrastructure built in the North Sea.

By and large the Norwegian licensing system has worked well. There are of course a number of alternative systems which could be used, and which are used by other countries. There is the US bidding system. One could use the international companies on a pure contract basis. The use of risk contracts is another possibility. We have found the existing system useful for our particular situation, because we have the benefit of applying on our continental shelf the many different geological philosophies applied by the various oil companies, while at the same time we are able through direct Statoil participaton to control the operations.

OBJECTIVES

In developing and carrying out our petroleum policies, a number of objectives have to be taken into consideration.

I think we could safely say that the overriding objective is to keep national control of these valuable resources on our Continental Shelf. The way of exercising national control has definitely changed over time. In the first phase of the development of the Norwegian Continental Shelf—when nobody really had any idea about the magnitude of these resources—the authorities tended to look upon the petroleum as any other natural resource, which according to our general philosophy is exploited by private companies, if that is feasible. The Government therefore looked upon itself primarily as a concessionary and regulatory agency. It did not intend to let the international oil industry do whatever it liked to do. Their activities were strictly regulated by the relevant government agencies, but the government was not an active partner in exploration or production. But fairly soon it became quite clear that the North Sea oil was going to be very important, from a national economic as well as from many other perspectives. If this resource was to be exploited, it was going to have a substantial impact on the total Norwegian society. Here we have to remember that we are not talking about an economy consisting of one hundred or two hundred million people, but a country of four million people.

The conclusion was drawn that only by active participation in the petroleum activities—in the concrete practical operations, as owners and operators—could the Government gain the experience and develop the in-house technological expertise necessary to maintain national control of these activities. This is background for the establishment of Statoil, and for the increasingly wider role this company is playing.

This development is also in line with the development we see in many other oil-and-gas-producing countries. An important point in this connection is the fact that very advanced technology is needed for the further development of these resources. This fact calls for cooperation between national as well as international oil companies and the authorities of the various countries. We will probably also in the years to come see a much closer contact between the authorities in different countries in questions concerning the exploitation of the continental shelves, and the marketing of oil and gas. The national oil companies will no doubt play an ever-increasing role in these intergovernmental relationships.

Another important objective of the Norwegian petroleum activities relates to level of activity. A moderate rate of extraction has been an established policy for quite some time. No production ceilings are used, but ninety million tons a year (1.8 million barrels per day) of oil equivalents (oil and gas) have been mentioned as an illustrative figure of what a moderate level might be. Through this policy we want to avoid becoming too dependent on one type of industrial activity. Experience has taught us, however, that development of oil and gas fields in the North Sea is more time consuming than originally estimated, and—as I have mentioned previously—production from existing fields will not reach more than sixty-five million tons per year.

The level of production is not regulated through any kind of production control, but through the licensing system, which means that long-term, detailed, aggregate production-planning is difficult. The total long-term production level is regulated through the number of blocks licensed.

Another objective of the Norwegian oil and gas policy is that the development of these resources shall not be detrimental to industries, like fisheries, or to the traditional Norwegian way of life. This requires participation from many different groups in the formulation and carrying out of these policies. Other relevant ministries, the fishermen's organizations, relevant local governments and the like provide input when important decisions are taken. And decisions on which blocks to select for production licensing have to be approved by Parliament. All this does not mean that it is a goal of the government to preserve Norwegian society as it is. Living societies are always changing. But it is an aim that this change should take place in an orderly and well-planned fashion, and not as a result of letting the economic forces loose, so to speak.

As regards petroleum activities off the coast of Middle and Northern Norway, where one is more dependent on fisheries than other parts of the country, the fishing interests have played an important role in the decision-making process. The selection of blocks to be leased has definitely been made with the fisheries in mind.

Oil should be used to make Norway a qualitatively better society. This objective is reached through a variety of measures. The Government tries to use the oil as a basis for further industrial development, thereby strengthening the country's industrial base. This is particularly important for developing economically weaker areas of the country. A large part of Norwegian industy has, in a relatively short period of time, been able to develop its technology into petroleum-relevant technology. This has been particularly important during a period of decline in shipping, shipbuilding and activities related to these industries.

It is a stated policy that Norwegian goods and services should be used on the Norwegian Continental Shelf as far as they are competitive with regards to quality, service, schedule of delivery and price. This means that Norwegian companies have a chance to compete in a business where contracts often are made in accordance with habit. A contractor often buys goods from companies and uses subcontractors with whom he is familiar and has had good experience. But this policy also means that Norwegian producers and contractors have to be competitive in order to get the orders desired. Statoil—through its operational role in some licenses, and its ownership role, where it is not an operator—has a special responsibility for seeing that these procurement policies are implemented.

When selecting the blocks for the fourth round of licensing, the Government made it clear that one of the elements to be considered in the granting of blocks would be the applicant company's willingness to cooperate with Norwegian industry in industrial projects which would contribute to a development of the technological level of that particular industry. This could be in petroleum or any other field. Many of the international oil companies have diversified their activities, and could thereby have a beneficial effect also in fields far away from petroleum. This is another element in the Government's effort to use oil to improve the overall industrial level in Norway. How great an effect this particular measure will have it is too early to say.

Energy production and consumption, like other industrial activities, often take place in potential conflict with the desire to have a clean environment. Norway is no exception, which was clearly demonstrated during the 1977 blow-out at the Ekofiskfield. Clean air and unspoiled nature have traditionally been highly valued in Norway, which is one of the most sparsely populated countries in Europe. In order to take care of—among other questions—the environment aspects of the North Sea oil development, a Ministry of the Environment has been established. This ministry has an important voice in all questions where there is the possibility of environment hazards.

Great emphasis has been put on developing an emergency system to take care of oil spills, and the Government is actively involved in developing the necessary equipment to take care of them, which we by stringent safety measures hope to avoid, but have to prepare for. Some of the objectives of Norwegian oil and gas policies are clearly defined. Others are more implicit. But these activities have no doubt received so much attention that the objectives are better developed than public policies in many other areas.

Looking back over the ten-to-fifteen years these activities have taken place, one can put the following question: Have the objectives and the policies changed over time? Basically, the answer is probably no. Today we know much more about the consequences of the activities of the petroleum industry on the Norwegian Continental Shelf than we did just a few years ago. One example of this is that the possibility for overheating the economy of the country has turned out to be much smaller than previously expected. This greater amount of experience has provided for a possible reordering of priorities between the various objectives, and an adjustment of measures to attain them. But the objectives as such have probably not been changed.

When analysing the key factors behind Norway's petroleum policies, we will have to look at a number of factors. One is the country's energy situation. As long as Norway has been an industrial country, our energy-supply situation has always been relatively favorable. The abundance of hydro-power formed the basis for the development of a sizable power-intensive industry. The hydro-power has been able to supply all the electricity the country needs, at least up till now. And this renewable resource even today supplies about half of the country's energy needs. Energy imports have therefore not been an important item on the balance of payments list. The country has not been in an energy-supply situation requiring fast development of the oil and gas resources on the Continental Shelf. With a domestic consumption of about eight million tons of oil and no natural gas per year (160,000 barrels per day), Norway is already a net exporter of hydrocarbons, and will be increasigly so in the years to come.

The important role played by fisheries in certain regions of the country is another element which has contributed to the policy of moderate development. So is the value put on a clean environment.

When it comes to the question of national control, and the role the Government played in the concrete operations themselves, we have to remember that Norway is a country where the state traditionally has played an important role in the economy, through full-or-part ownership in business enterprises in various fields. State participation in petroleum activities as such is therefore not a controversial question.

CONCLUSION

I have here only been able to touch upon a few of what I consider to be the most important aspects of Norwegian oil and gas policies. I have not gone into the economic magnitude of these operations, which for a country of our size is considerable. And I have not gone into the international implications of Norway as an oil-exporting country. But I think I have touched upon the aspects which are most relevant in this context.

Commentaries

DON KASH

US Geological Survey
United States

All the points Martin Belsky made can be summarized in saying that the Outer Continental Shelf Lands Act Amendments represent an effort to put together an OCS oil and gas policy in this country, and represent the same kind of outcome one would anticipate for any highly controversial policy.

The Act is a series of contradictions, an effort to satisfy conflicting needs, and, in some instances, it is possible to read sections of the Act which appear to be in conflict. In some instances it is clear that compromise was imagined to be possible either by finding a way to respond to both interests or by ducking the issue completely. This process of accommodation represents precisely the character of OCS policy in this country. It is a cauldron, a collection of conflicting desires and concerns. Goals are unclear. The Act seeks to develop oil and gas as rapidly as possible, while providing maximum protection for the environment, maximum opportunity for every conceivable interest to participate at every possible level in the decision-making and implementation process, and some payoff to the states in dollar terms through various indirect means.

In total, the Amendments represent what one would expect for a major area of public policy. We might compare that Act and the way in which we manage OCS policy in this country with what we learn from the two papers from Norway and United Kingdom.

As a Federal bureaucrat responsible for implementing OCS oil and gas policy, I am burning with envy at what they do in the UK and Norway. I can't think of anything more improbable in the United States with regard to OCS oil and gas policy than that government would turn over to a set of bureaucrats the opportunity to negotiate licenses and their content.

The first time I went to the United Kingdom, when they were developing oil and gas policy, I had a discussion with people in the Department of Trade and Industry. When asked about their regulations, this very debonair Englishman said, "Oh, we just go down and talk with the chaps, sort of discuss what makes good sense, and then we go ahead."

Now compare that with what we do in this country. We don't negotiate, at least in that way. We have highly formal, highly detailed, highly bureaucratic rules and procedures. Wherever possible we have infinite detail. We move from one straightjacket to another with regard to details. We formally give every interest an opportunity to participate in every action, no matter how detailed. Although people complain on the one hand about detailed regulations, what they really complain about are detailed regulations they don't like. They insist, also, on having detailed regulations to insure that they get what they like.

In substance, what we have with regard to policy in this country is a collection of infinitely greater details concerning procedures and substance. Now that leads to some interesting things. We have a requirement to process plans of exploration within thirty days. We also have another Act—the Coastal Zone Management Act—which says we are not allowed to do it in thirty days, so it requires creativity on the part of bureaucrats like myself. We work out all kinds of interesting mechanisms so that we can process the plan in thirty days and *not* process the plan in thirty days, both at the same time.

With the present energy difficulties in this country, there is a very real commitment to rapid development of the OCS while at the same time there are very real commitments to protecting the interests of the states and the environmental interest groups, as well as insuring broad participation in decisions. All of that creates a situation in which we cannot move with anything like the speed that either the Norwegians or the UK have in the development of their resources.

CHARLES McPHERSON

Continental Oil
United States

The preceding chapters provide an excellent introduction to the workings of and the outstanding issues associated with OCS oil and gas leasing policies in the three countries. I would like to elaborate on one important aspect of each of these leasing policies: namely, the mechanism used to allocate leases or licenses among companies.*

In discussing the relative merits of different allocative mechanisms, it is useful to keep in mind two "yardsticks" that appear common to all of the diverse goals of leasing policy: (1) that costly conflicts between the host government and the licensee (or more generally among all interested

*These are personal views and do not necessarily represent Conoco Inc.

parties) should be kept at a minimum; and (2) that pursuit of OCS goals should not grossly violate the requirements of efficient resource development. These are not entirely independent yardsticks: conflict usually wastes resources, and policies which waste society's resources will surely lead to conflict.

Martin Belsky made clear the Congressional concern in the United States that conflict and uncertainty be reduced, and, using two examples, held out some hope that such an improvement might be forthcoming under the new legislation—at least in the area of environmental reviews and lawsuits. But in the area of lease allocation, I suspect that OCS Lands Act Amendments legislation *may* have set the stage for increased conflict, and for less efficient resource development. I stress "may," because a lot will depend upon the skill with which the Interior and Energy Departments experiment with alternative leasing systems.

As you will recall, the legislation requires a five-year period of experimentation. Twenty to sixty percent of new leases are to be awarded using some system other than the traditional cash-bonus-bid/fixed-royalty system.

Measured against our two yardsticks, the cash-bonus bid is an excellent allocative device. Its simplicity, plus the fact that it transfers to the government the maximum surplus value or economic rent expected from lease exploration and development, combine to minimize the occasions for government/company dispute. At the same time, because the bonus is a sunk cost to the company winning the lease, it can have no distorting effect on exploration, development or production decisions; thus efficiency is safeguarded.

With one possible and intriguing exception, the alternative bidding systems now proposed for us on the OCS do not share the advantages of the cash bonus bid.

Several alternatives rely heavily on royalties. Royalties raise the costs of oil and gas development to the companies without affecting the "true" cost of that development, that is, labor, materials, equipment and the like. This can lead to situations where a company—acting in its own commercial interest—will be reluctant to develop a reserve or to continue production from a reservoir, despite an excess of the value of production over its "true" cost. Where royalty bidding is used to allocate leases, royalties will be much larger than conventional royalties, and these adverse efficiency effects may become serious. Further, reluctance to develop discoveries and premature abandonments, regardless of the commercial sense of such decisions, clearly will create occasions for conflict among government, companies, and interested public groups.

Sliding scale royalties, which have been used in several lease sales since passage of the Lands Acts Amendments, can bring their own special problems, increasing the potential for conflict and misallocation of resources.

As for work commitment bids, the best determinant of the size of a work program is not what it takes to win a lease but rather the explorationist's assessment of tract potential and geological character, as amended by the results of early seismic work and drilling. More often than not, the former consideration (winning a lease) will result in larger work programs than the latter (geological merit), with consequent over-allocation of scarce resources to the lease.

Allocation of leases on the basis of work commitment bid may precipitate government/company disputes, not only over the ultimate appropriateness of the work program, but also over the division of revenues. This is because the work bid, while likely to go beyond an economically justifiable level, will in all probability stop short of transferring to the government the full rent expected from lease development. In both the UK and Norway, this sort of situation was at least in part responsible for the introduction of supplemental taxes—the Petroleum Revenue Tax in the UK and the Special Tax in Norway. Renegotiations of this kind can result in expensive delays and discourage investors. Pressures to "Buy British" or "Buy Norwegian," and most recently the licensing criterion mentioned by Einar Risa—namely, the willingness of oil companies to undertake joint ventures with Norwegian industry outside their traditional areas of cooperation—may reflect the same "rent-capturing" phenomenon and place the same strain on investor interest and on company-government relationship.

We should be cautious in the US in how we assess the North Sea models in regard to work-commitment bidding. It seems to me that the rapid rate of offshore development in the UK after the first discoveries had much more to do with the rate of licensing, initially favorable terms, and experienced success, than with work commitments. Further, we really are not observing work-commitment bidding in the North Sea in the sense that it is proposed for the United States. In practice, both the British and Norwegian governments rely on more than one bid variable: beyond the work commitment, government negotiators look at company financial strengths; commitment to cooperate with local investors and businesses; purchases of local goods and services; technical performance record; willingness to provide for state participation; and many more. This means that companies can submit work programs approximating what is economically justifiable, —without excess commitment—and still expect to be in the running for the lease award.

Multiple criteria and a negotiated award may be preferable to reliance on a single bid variable where use of that variable creates the potential for trouble during the lease development period. However, as Don Kash has emphasized, the degree of discretionary authority allowed government negotiators in the UK and Norway is unthinkable in the present US leasing framework. Where a single bid variable measures up well against the yardsticks of efficiency and minimum potential for company/government disputes, the US system does have the substantial additional advantage of simplicity and clarity.

There is one alternative leasing procedure proposed for the United States which I personally find quite appealing: the net profit share bid. Our company does not have a position on the procedure, but to my mind it has considerable possibilities.

Under the net profit share system, the lease would be allocated to the company or companies offering the highest share of lease profit to the government, where profit is measured net of some minimum return on the company's investment. The minimum return would be established in advance of the sale. A company's bid would then reflect the financial premium which it thought necessary to compensate for risk of lease exploration and development.

Implementation of this system is probably not as difficult as it may sound. In fact, it could be fairly simple, based on cost data already compiled by the companies as a matter of routine, and routinely audited.

The system should stand up well against our yardsticks: the government could expect to siphon off a high percentage of the economic rent associated with development of a license, thus minimizing one possible source of conflict; and because the bid variable applies only after recovery of some minimum financial return, the potential for distortion of investment decisions is probably quite small, making for efficient resource development and further reducing the potential for conflict.

Net profit share systems may have one advantage relative to cash bonus bidding. I suspect that governments are going to have an increasingly difficult time letting alone the income from offshore leases acquired by bonuses which were paid when income expectations were considerably less than income actually realized. The tension that is bound to result would be lessened under a profit share system where the government's "take" depends on actual rather than anticipated profits.

These comments barely touch the surface of a complicated and contentious issue. The lease-allocation mechanism is bound to be a part of any debate over leasing policy for some time yet.

PAUL STANG

Department of Interior
United States

I want to cover a couple of subjects. First is the five-year leasing program for OCS oil and gas. The preceding chapters represent opposite ends of a spectrum. At one, the Government has full discretion to negotiate whatever terms it wants with industry. At the other, the Government employs a very highly regulated approach.

Government tries to lay down everything in precise detail so we know what will happen and how this control will take effect. This approach certainly is a trend in our country. Congress has written laws which are more and more specific, more and more detailed. In turn, US federal bureaucrats have the job of preparing implementing regulations that are ever more exacting.

We have found complaints and problems with both systems. Perhaps we are missing a middle ground. There are other ways to meet national objectives than either extensive regulation or total discretion between a national government or national oil company and a private interest. A few that come to mind are under the general category of "dis-incentives." These "dis-incentives" are results produced which are entirely different from what the regulator intended. Or it may be what the regulators intend, but with side effects producing very undesirable results. Before going over these, let's look at some "dis-incentives" which are being built into recent legislation and regulations.

"Diligence" is one example. It has become an increasingly popular concept. Diligence provisions are often employed to assure that private interests develop public resources, whether they be ocean modules or oil and gas, according to a certain set of criteria to limit or eliminate private gain from speculation. Speculation is often viewed as intrinsically bad. This view pervades some of our legislation. True, some speculation is bad and detrimental to the national interest; but there are other aspects of speculation which are an efficient way to allocate and utilize a resource. The "all-speculation-is-bad" concept, when it becomes law in the form of an unsophisticated diligence provision, can result in a net cost to the nation.

Another "dis-incentive" is the "work plan" or "work program." There are some beneficial aspects of work plans and work programs, but the tendency is to lock-in a company or private interest to a set operational scenario at a very early stage in the process,when little information is available. Once locked in it is difficult to change, even though adjustments may be warranted, given changes in economic circumstances or in national goals.

Another "dis-incentive" is the royalty, a high royalty payment. Suppose the government issues you a lease for which you pay "X" dollars up front, which is, let's say, a rather small amount relative to the total value of the public resource which you are to develop. The lease calls for, let's say, a sixty percent royalty of the production of the oil. Well, it turns out that in oil production the rate of recovery could be rather fast in the initial stages of the depletion of the reservoir. However, as you deplete the reservoir, it becomes more and more expensive to take out the oil.

Since you have to pay the government a high percentage of the production (sixty percent), you stop production earlier than if you had a lower royalty. You stop producing oil when your costs rise to the point equal to the value of the forty percent of the oil you get to keep. To produce beyond that point is economically inefficient. The result—a lot of oil is left in the ground which could otherwise be extracted.

Another "dis-incentive" mentioned previously is the "Buy US," "Buy Norway," concept. It seems to have some benefits, but, again, distorts the market system.

Now, let's look at some "incentives." Profit-sharing is provided for in the 1977 amendments to the OCS Lands Act. Profit-sharing can be an incentive for optimal utilization of resources. It is not a production requirement approach, but an incentive approach.

Another incentive which could be explored more is use of performance standards rather than production rates. Another is the good old cash award. Think of how we as employers and we as employees interact with each other. We aren't generally a production standard. We are on an incentive system. We do a good job and we get paid accordingly. That concept can be expanded to the way government does business. Coastal states have found it an incentive to join the coastal zone management program. They are not required to, but the incentives are there.

Another interesting incentive is full information disclosure. The National Environmental Protection Act (NEPA) really in many ways is a non-regulatory approach which is an incentive to accomplish some national interest. Under NEPA, one must write a rather lengthy statement, but as a result of information disclosure one is *not* obliged by NEPA to do or not to do a particular set of activities. One is only obliged by NEPA to make full disclosure. More environmentally sound projects result either because one understands the ill effects of one's original concept and voluntarily changes it, or because one is sued by an environmental interest group, or fears one might be. The public concern that is produced from the information is the incentive to change.

The bottom line is that there are other approaches than the regulatory approach. The regulatory approach limits creativity and discretion in areas where discretion and creativity can help promote innovation toward a more efficient system, both for government and private industry.

I would like to mention another specific subject, that of Coastal Zone Management (CZM). It is an interesting time right now for CZM with respect to the production of oil and gas and the whole leasing process.

This gets into some particular US interests. The CZM Act became law in 1972. It has two "federal consistency requirements" which are of interest here. One says that "Federal activities which directly affect the coastal zone of the state must be consistent"—and then a caveat—"to the maximum extent practicable."

The second one places a more severe consistency restraint on people who get licenses and permits—for instance, oil companies who must get a permit from the USGS.* This more stringent requirement is that if activities performed "effect the land and water uses of the coastal zone," then they must be fully consistent. The coastal state has a "veto" over the permitted activity if the state determines that it is "inconsistent with their program."

The Department of Interior (DOI) had a difference with the Department of Commerce (DOC) as to whether or not Interior's OCS oil and gas or leasing activities are subject to consistency. In response to a DOI/DOC request, the Justice Department said in April of 1979 that the consistency provisions do apply to Interior's pre-lease activities. However, Justice went on to say that the language that Congress wrote on the consistency of federal activities versus the consistency of licenses and permits should not be blurred. There are distinctions there, and they should be maintained. What is especially timely about the Justice opinion is that the Interior Department is engaged in OCS pre-lease activities for the upcoming Sale 48 off of California and Sale 42 on Georges Bank. These consistency issues are right at the forefront.

We will see in days to come how they evolve and are worked out. It is curious to note that the law was enacted in 1972 and yet we have not worked out what that law means and how it will be implemented.

*US Geological Survey

FIVE

Fisheries Management

Lars Vidaeus
New England Regional Fishery Management Council
(presently with World Bank)

CHAPTER 11

Marine Fisheries Management in Canada: Policy Objectives and Development Constraints

PARZIVAL COPES

Simon Frazer University
Canada

Canada's marine fisheries policy in the era of the two-hundred-mile limit reflects the country's position as a developed coastal state with a severe regional economic problem in the part of the country most dependent on the fishery. Policy objectives have focused on utilizing fish stocks placed under Canada's jurisdiction to make a maximum contribution to the enhancement of employment opportunities and income levels in coastal communities. Mindful of its good reputation as a member of the international community, Canada is nevertheless anxious to retain harmonious relationships with distant-water fishing nations. Its external policy is to make stocks available to foreign fleets on reasonable terms where they cannot be utilized effectively by domestic operators.

Under Canada's federal constitution, authority in the area of fisheries resource management is assigned entirely to the central government, which is therefore able to integrate singlehandedly the external and domestic aspects of fisheries policy. The provinces, however, have jurisdiction in a number of areas impinging on fisheries development generally. The Atlantic Coast provinces, in particular, have taken a keen interest in fisheries matters because of their local political importance. The need has been recognized for federal-provincial cooperation and coordination of policy and administration. In recent years an elaborate complex of consultative arrangements has developed between the two levels of government, in some instances extending to the fishing industry and to fishermen's organizations. Less elaborate consultative arrangements have been set up to involve the Pacific Coast province of British Columbia.

Much of the federal-provincial consultation is of an ad hoc

character, but it has been on a broad scale. The federal government has been careful, however, to retain firm control of fisheries resource management per se and to avoid fragmentation of decision-making powers. The enhanced prospects for the fisheries that have come with the two-hundred-mile limit have whetted the appetite of the provinces for greater authority and control over fisheries. The platform of the federal government that came to power in the 1979 election included a promise to share out greater powers over fisheries to the provinces. This commitment had yet to be translated into specific measures, so the impact on content and coordination of fisheries policy cannot yet be determined.

Internally, Canada's fisheries policy in recent decades has moved away increasingly from simple biological management to bio-economic and socio-economic considerations. Restricted-entry regulations are being introduced in all fisheries, and buy-back programs have been applied in a number of areas. Effort is being deliberately limited to meet the criteria for an (as-yet-ill-defined) optional sustainable yield. The most severe constraints or attempts to rationalize fishing operations have come from the pressures of unemployment and underemployment in the Atlantic Provinces. There is constant pressure to absorb surplus labor in the fishing industry. This is biasing fisheries administration in favor of more labor-intensive and less productive inshore operations, resulting in a lower income potential per employed worker and a continuing need for subsidization of the industry.

Externally, Canada's policy has been to offer interested distant-water fishing nations access to surplus stocks against a variety of fisheries considerations of benefit to Canada. The latter have included access for Canadian fish products to foreign markets, processing of foreign-caught fish in Canadian plants, utilization of Canadian port and repair facilities by foreign fleets, over-the-side sales of Canadian-caught fish to foreign factory-vessels, foreign license-fee contributions to Canadian management, recognition of Canadian management objectives in NAFO, the Northwest Altantic Fisheries Organization, and abstention from high-seas salmon fishing.

The two-hundred-mile limit has complicated Canada's important fisheries relationships with the United States. There are inevitable disputes over location of two-hundred-mile zone boundaries. Fishing patterns are having to be rearranged because fishermen in each country are losing access to areas off the neighboring country's coast. In the face of US determination to expand domestic use of coastal stocks, Canada may have to find new markets for much of its volume of fish exports to the US. Both countries appear to recognize the need for joint management of important transboundary stocks. The prospects for cooperation and compromise remain promising.

CHAPTER 12

EEC Fisheries Management Policy

JOHN FARNELL

International Fisheries Division
Commission of the European Communities

HISTORY OF THE EEC

There are often misunderstandings about what the European Economic Community (EEC) is. Since January 1973 the Community has consisted of nine countries: Belgium, Denmark, Germany, France, Ireland, Italy, Luxembourg (which has little interest in fisheries policy), the Netherlands and the United Kingdom. Greece will become a member of the Community at the beginning of 1981, and Spain and Portugal are in the process of negotiating their accession.

The Treaty of Rome, under which the Community was founded in 1958, laid down a number of areas of economic activity in which the Community should achieve a Common Market and progressively approximate the economic policies of its Member States. Under Article 38 of this Treaty, the Common Market was extended to agricultural products, including those of fisheries, and it also laid down that operation and development of a Common Market in agricultural products must be accompanied by the establishment of a common agricultural policy and, therefore, by extension, a common fisheries policy.

The Treaty also established the institutions which would be responsible for achieving the goals of the Community. These institutions are, first of all, a parliamentary assembly, which in 1979 was elected by universal suffrage for the first time; a Council of Ministers, composed of representatives of Member State governments; the Commission (for which I work), which is an independent body composed of members nominated by governments but not responsible to them; and finally, a Court of Justice, which interprets community law and exists to guarantee respect of the application of the Treaty.

The Commission formulates proposals for legislation necessary to achieve Community policies. It presents these to the Council for a final political decision, and is responsible for the operation of these policies once they have been decided upon.

EEC FISHERIES PRINCIPLES

In 1970 the Community adopted regulations establishing a Common Market in fisheries products and a common structural policy for the fishing industry. In that latter policy, the structural policy, the principle of equal access for Community fishermen to the waters of any Member State was made clear for the first time.

The next step was in 1972, when under the Treaty of Accession to the EEC by Denmark, Ireland and the UK a ten-year derogation to this rule of equal access was made, and, at the same time, the Council undertook to determine the conditions for fishing in Community waters before January of 1979.

The third basic step was taken in November 1976, when the Community agreed on three further principles:

> First, to the extension to two hundred miles of all the Community Member States fisheries zones;
>
> Second, to the transfer to the Community of exclusive competence for the negotiation of EEC fishery agreements with non-Community countries (these negotiations are conducted by the Commission); and
>
> Thirdly, the Community decided upon certain guiding principles for a future Community policy on the management and conservation of fisheries resources.

The most important of these principles were the commitment to the development of fisheries in certain outlying regions of the Community particularly dependent on fishing, and an undertaking by the Member States to adopt conservation measures for their waters only in the absence of Community measures.

Since the end of 1976 the Community has been working towards the development of a fisheries policy. It has made considerable progress in its external fisheries policy. We have negotiated framework fisheries agreements with all our principal fisheries partners in the North Atlantic, as well as one African country, and have completed eight different negotiations in all.

On the internal policy side, however, the picture is rather different. Although the Commission has presented a package of proposals for internal fisheries policy, agreement has not yet been reached by the Council, that is, by the Member Governments. This is because there are differences over the management and allocation of fisheries resources.

While some significant conservation measures have been adopted by the Council, and while the debate on the long-term policy continues, we have not yet adopted a fully integrated policy.

EEC FISHERIES PROCEDURES

The Commission, in its role as the coordinator of EEC fisheries policies and representative of the Community in its external dealings, is in constant contact with the national and international organizations concerned with fisheries. Representatives of the Member States' fisheries administrations attend regular meetings of working groups in Brussels, at which the Commission is represented, in order to prepare the decisions of ministers, or to advise the Commission during negotiations with non-Community countries. To give you an idea of how frequent these meetings are, in the case of our external fisheries policy group there is a weekly meeting of national fishery officials and Commission representatives, and, on the internal side, because there is perhaps less urgency, the meetings are held more intermittently, perhaps every other month. There is also a Consultative Committee of representatives of the fisheries industry, which has been in existence since 1970, whose task is to comment on policy proposals and to offer technical advice to the Commission. The Commission also has two kinds of contact with the scientific community. Firstly, there are direct relations with the International Council for the Exploration of the Sea (ICES), insofar as the Commission has asked this organization for advice on Community fish stocks. Secondly, we convene scientists from the Member States' national fisheries laboratories in order to obtain a "second opinion" on matters of particular interest.

While the lack of agreement on an integrated EEC fisheries policy means that our practical experience in fisheries management is rather more limited than that of other North Atlantic coastal states, a number of policy initiatives which have been taken in the field of conservation measures and structure policies could be of general interest.

EEC FISHERIES POLICY

Our fisheries management policy in the Community is essentially one of response to a particular situation, rather than a policy which has been founded on a set of guiding principles. It began as a reaction to events, although with the accumulation of experience and as the broader implications of our fisheries policies have become more clear, there has been much more thought given recently to the fundamental basis of our policy for the future. In very brief terms, the situation of the Community in the mid-1970s, which prompted an early development of the common fisheries policy, was as follows:

First and foremost, the scientists were unanimous that fisheries resources within what is now the Community zone were substantially depleted. The inability of international organizations to prevent over-fishing had resulted in dramatically reduced catches of traditional species for Community fishermen. A major part of that over-exploitation could be attributable to the very big buildup of foreign fishing. In 1976 foreign fleets fished about a third of the total catch taken in Community waters, and the scale and methods of the fishing, to put it quite bluntly, meant that conservation action had become an absolute priority.

At the same time, the EEC fishing fleets had been growing as a result of uncoordinated national investment policy in the previous decade. For example, there had been a buildup of the Danish fishing fleet for catching fish for reduction purposes, and a similar buildup of the Scottish trawler fleet as a part of British regional policy. In fact the Community had, and still has, fishing capacity far exceeding its available resources. These problems were intensified by rapid shrinking of fishing possibilities outside Community waters as others extended their fisheries limits. I should recall that the UK fishing fleet fished "in the good old days" about thirty percent of its total catch outside what are now EEC waters and the German fleet as much as eighty percent. We were faced with the problem of having to preserve, on the one hand, such fishing as was available outside the Community by negotiating agreements and to consider, on the other hand, how the largely redundant long-distance fishing fleet could be employed in Community waters.

The final element in our package of problems was the importance of fishing as a motor for regional development in certain parts of the Community. Although the conservation crisis might in principle require a severe limitation of fisheries activity for the EEC as a whole, in areas such as Ireland, Greenland, and Scotland, fisheries were seen as ideally suited to regions which were economically depressed but disposed of extensive fisheries resources. This political commitment to regional development is still a major consideration in our thinking on fisheries policy.

MEETING THE PROBLEMS

To take conservation of depleted resources first, our approach has been rather different to that adopted by the United States and Canada, which are countries concerned to establish the optimum yield for resources which are, with some exceptions, surplus to their domestic fishing requirements. The Community has been faced with a chronic problem of excess fishing capacity, and has therefore had to take emergency measures to conserve some fish stocks, such as, completely banning the fishing for herring in the North Sea and west of Scotland, even though 300,000 tons were fished from those stocks only four or five years ago. We have had to fix TAC's Total Allowable Catches for stocks which were in better shape than the herring at the highest level acceptable to scientists, given our very difficult structural problems. Even so, we have taken an enormous drop in catches of groundfish. Average catches of these species in the Community were one and a quarter million tons in the four years up to 1976, and they are now only 800,000 tons per annum; that is a drop of one-third.

So the EEC has made some progress in its decisions on conservation. But we have also taken non-quantitative management measures, and I would like to mention one or two of those. For instance, certain areas of the North Sea have been closed to fishing for industrial purposes, and the use of herring for reduction purposes has been completely banned. Mesh-size and by-catch regulations have also been tightened up in comparison with the situation which was permitted under international organizations before the extension of fisheries jurisdiction.

Potentially the most far-reaching restriction of fishing activity now under consideration by the EEC is the use of fishing plans. This attempt to match fishing capacity to available resources has been attempted elsewhere—I am thinking particularly of Canada with respect to foreign fishing—but never on the scale on which it might have to be used within the Community. Now, although in theory fishing plans are an ideal mechanism for limiting fishing effort and affording greater protection to certain coastal communities which need to be given special consideration, the concept raises problems for the EEC for two reasons. The first is legal. As I have already indicated, the principle of non-discrimination between our Member States and equal access to all EEC waters is enshrined in Community law. Secondly, there are real operational difficulties about fishing plans. We are dealing with several national fleets, which have different kinds of vessels and techniques of fishing; this complicates the assessment of catch capability. There is also an enormous mix of fish species in the North Sea which means that the fishing plan to protect one species will inevitably have an impact on several other fisheries. For these reasons it is likely that fishing plans will be used only on an experimental basis in the near future.

How far has the Community come in deciding upon a system of catch allocations between Member States? In a formal sense, we have not yet agreed on the allocations between member countries owing to considerable differences between the United Kingdom and the eight other Member States as to the criteria to be employed in making these allocations. The approach which is being proposed by the Commission, and which is supported by the other eight Member States, has been to base allocations on historical performance in each fishery, with two kinds of adjustment. The first is a weighing of these allocations where the stock concerned is of interest to a coastal community which has been designated as particularly dependent on fishing; the second is the need to compensate those Member States which have suffered losses of fishing in external waters. We have not agreed on the proposed allocations because there has in effect been a veto exercised by the United Kingdom, which has put forward two counter-arguments: first, that the contribution to overall EEC fisheries resources made by the United Kingdom should be taken as a further criterion for allocation of resources; second, that historical performance relating to "EEC" fish stocks has been distorted by the fact that the United Kingdom has fished extensively in other countries' waters in the past.

The Council has decided, pending agreement on the formal Commission proposals, that there should be a standstill for fishing activity within the Community zone. Catches have been frozen at the levels of 1976; this is, of course, besides the ban on herring fishing.

STRUCTURAL POLICY CONSIDERATIONS

The adaptation of industrial structures is an important part of our preoccupations at the moment. I have already mentioned one step we have taken in this direction, which is a limitation on what we call "industrial fishing", or fishing for reduction purposes. In the present state of EEC fisheries resources, this type of fishing will probably only be able to develop through the exploitation of new species which have not been exploited so far, such as blue whiting or horse mackerel. This fishing will have to be redirected out of the central area of the North Sea to the fringe of the Community fisheries zone, where these unexploited stocks mainly exist.

We have also taken certain measures—and have proposed others—to use financial incentives to effect long-term changes in the composition and structures of the EEC fishing fleet. Firstly, there has been a proposal to offer financial compensation for the temporary laying-off of vessels which normally fish stocks which are subject to conservation requirements. We foresee longer-term incentives to scrap vessels which are surplus to capacity requirements. The second direction we are taking is to contribute to the cost of construction or purchase of small-scale fishing vessels for coastal fisheries. We have had a scheme of this sort in force since last year, and we foresee that coastal fisheries is the one area open to future development.

Other developments in the Community's structural policy are still a matter for debate. Among the possibilities open to us are a harmonization of national-aid policies in the sector of fisheries such as we have already achieved in agriculture. This would mean that procedures would have to be set up for the approval of such aids by the Community, which would obviously depend upon their compatibility with our overall conservation objectives and our objectives for the development of the structure of the industry. This procedure would make it impossible for a Member State to subsidize construction of vessels which have no real hope of fishing at their full capacity in EEC waters.

Of course this area raises difficult conceptual problems for a Community which was founded upon an essentially non-interventionist "laissez-faire" philosophy. How far should the Community go in directing the expansion or construction of fishing fleets? Do development plans or market forces do the job of allocating resources and determining production methods most effectively? We do not have answers to these questions yet.

REGIONAL FISHERIES POLICY

I should now like to refer very briefly to a problem concerning regional policy in the fisheries context which I think may be of wider interest, namely, a conflict between the effectiveness of regional policy in achieving social objectives such as full employment, and the realization of other objectives, such as conservation and the balance between supply and demand for fish. With the increasing sophistication and efficiency of fisheries technology, those fishermen who benefit by the sort of financial incentives for coastal fishing to which I have just referred may increasingly put conservation requirements at risk. There seems to be an analogy here with a concept which may be familiar to economists: the famous Phillips Curve, which demonstrated that the more employment went down, the more inflation went up, and vice versa; (it does not seem to apply now, but it used to apply). In the fisheries context we are dealing with an apparent incompatibility between full employment and fisheries technology. You may choose one, or the other, and the fish will survive. But if you equip the entire potential labor force with the latest technology, the fish may very well disappear. The correct balance between these variables—which are at the same time political, technological, and biological—will be extremely difficult to find.

CONCLUSION

Two further points should influence our thinking about domestic fisheries policies, and change our rather standard assumptions about how things are going to develop from here.

The first factor is the possible effect of an "energy crisis" or a shortfall in energy on the fisheries sector. The standard fisheries technology of today is, I would maintain, highly energy-dependent. As energy becomes more costly, there may very well be a shift to more labor-intensive kinds of fishing. That could very well turn out to be a major "break" for fisheries officials who are trying to find a balanced solution to the problem of regional fisheries development.

The second factor which may have to change our thinking about fisheries in the future is the change in the pattern of demand for the end-product of the fisheries industry. Demand for foodstuffs—whether produced by farmers or by fishermen—is highly inelastic in the high-income societies of the North Atlantic area. In the long run, fewer fisheries resources may be required by our industrialized societies, although in the developing world demand will be as urgent as it is now. In Western Europe, at least, we have found that there has been a steady decline in overall fishing consumption over the last ten years. The position of fish in relation to competing foods is another unpredictable variable. We are not able to predict any certain trends.

Among those current policies which may have to be reviewed in the light of future demand must be the development of a heavily-industrialized structure for fisheries within the North Atlantic area. The problem is, who is going to eat those fish, or rather who is going to *buy* them? I think there is a danger that our markets may become saturated. Of course there is a market in the Third World, but at what price? Will the price be high enough to sustain the income expectations of a highly-developed industry in the North Atlantic area? I leave you with that somber food for thought.

CHAPTER 13

Management of Living Marine Resources: Challenge of the Future

WILLIAM GORDON

National Oceanic and Atmospheric Administration
United States

INTRODUCTION

Until recently the US Government—like many other national governments—had limited or no authority over living marine resources beyond the territorial limits, except under legislation implementing US participation in international fishery agreements. Although there are exceptions, most international conservation efforts can best be described as creating regimes that prevented more serious resource declines than otherwise would have occurred. International agreements did not create a single management entity charged with the responsibility to make tough decisions nor—more importantly—to implement and enforce these decisions. As a result, the various international fisheries commissions were: (1) unable to ensure optimum harvesting of the resources; (2) ineffective in controlling or allocating catch and effort: (3) incapable of resolving major issues among competing groups of foreign nations: (4) unable to make timely management decisions required to achieve conservation; and (6) not constituted to deal with major social and economic considerations.[1]

During the early 1970s, three federal laws drastically changed the role and authority of the US Government in the management of living marine resources, and provided a basis for effective stewardship of most living marine resources within an expanded jurisdiction. These laws are the Marine Mammal Protection Act, the Endangered Species Act, and the Fishery Conservation and Management Act of 1976 (the FCMA). The first two Acts are administered jointly by the Secretaries of Commerce and Interior, and the third is primarily the responsibility of the Secretary of Commerce. It is the FCMA that this paper addresses, since it is the principal Act dealing with the marine fisheries of the United States and has been described as the most significant fishery legislation in our history. It provides a new institutional framework for managing and conserving about fifteen percent of the world's marine fishery resources. Prior to its passage, few other nations had established similar fishery zones. Many have since done so by legislation patterned after the FCMA. By now the two-hundred-mile extended fishery jurisdiction is generally accepted as customary international law. Implementation of the FCMA is the dominant factor in US marine fisheries policy at this time.[2]

The Act established, for the first time, a comprehensive system for managing fisheries in a Fishery Conservation Zone (FCZ) that extends seaward from three to two hundred nautical miles. The Act thus extended the legal jurisdiction of the United States from an area of about 545,000 square nautical miles to an area of over 2,000,000 square nautical miles. Although a major aim of the legislation was to curb foreign fishing off US coasts, the Act's management controls apply equally to domestic fishing. The policy brought into effect by the Act is to assure that fish are harvested responsibly in accordance with regionally developed plans. These are to be based on the best available scientific information and must meet specified national standards.

A key goal of the Act is attainment of what is termed the "optimum yield" of each fishery, a concept that denotes the amount of fish that will provide the greatest overall benefit to the United States, including both food production and recreational opportunities. The concept not only takes into account the maximum harvest that will permit a species to sustain itself, but also includes consideration of any relevant economic, social, political, or ecological factors.

The Act limits foreign fishing in the US two-hundred-mile zone to that portion of the optimum yield that will not be harvested by US vessels. Both the optimum yield and the amount to be made available for foreign fishing are determined by the eight Regional Fishery Management Councils established by the Act and approved by the Secretary of Commerce. The surplus is allocated among foreign nations by the Secretary of State in cooperation with the Secretary of Commerce. Permits to fish are issued to individual vessels by the United States, and enforcement is the responsibility of the US Coast Guard and the National Marine Fisheries Service, a part of the Department of Commerce.

Domestic Fishery Management Plans (FMPs) are drawn up by Regional Fishery Management Councils and approved and implemented by the Secretary of Commerce. To date, most FMPs have been prepared on a species-by-species basis, covering a specified geographical area.

Although the Councils are Federal instrumentalities, their membership is drawn primarily from nominations made by the governors of the states in each geographic coastal region of the country. Thus, the Councils' membership comprises a blend of state and federal officials and members of the public having knowledge of fishery matters, and environmental, conservation, and consumer interests. Each Council has a Scientific and Statistical Committee and Advisory Panels, which provide advice on each fishery management plan they prepare. The Advisory Panels are likewise composed of local individuals with knowledge in the fisheries involved. Along with the national standards, the Act sets forth requirements for each fishery management plan, providing the Councils with broad authority within the confines of these standards and requirements.

It is apparent that the Act's carefully constructed balance of local, state, regional, and federal participation in its implementation constitutes a unique form of governmental decision-making, juxtaposing local and regional expertise and sensitivities with national and international perspectives. As in any pioneer effort, there are many initial challenges. This paper addresses some of the more significant problems which surfaced in the first three years of operation.

This paper is in three parts. The first part summarizes the Fishery Conservation and Management Act of 1976 and other applicable laws. The second part addresses some of the principal problems encountered in the formative years of implementation. The final part presents some conceptual aspects with which the national management must cope in future years, if the potential and promise of extended fisheries jurisdiction are to be realized.

FISHERY CONSERVATION AND MANAGEMENT ACT OF 1976

I. The FCMA

A. *Findings - Act Recognizes:*

1. Fish are renewable national resources which contribute to food supply, economy, and health of the nation and provide recreational opportunities.

2. Increased fishing pressure and inadequate conservation have contributed to reduction in numbers of fish to a point where survival of some stocks is threatened.

3. International fishery agreements have not been effective in

preventing overfishing.

4. A national program for conservation and management of fishery resources is necessary.

5. A national program for development of fisheries is necessary.

B. *Policy*

1. Maintains without change existing territorial or other ocean jurisdiction of the United States.

2. Authorizes no impediment to, or interference with, recognized legitimate uses of high seas.

3. Assures use of best scientific information available.

4. Involves and is responsive to the needs of affected states and citizens.

5. Permits foreign fishing to continue.

6. Supports and encourages continued active US efforts to obtain Law of the Sea treaty.

C. *Fishery Conservation Zone (Sec. 101)*

The FCZ is a zone contiguous to the territorial sea of the United States, the inner boundary of which is a line coterminous with the seaward boundary of each of the coastal states, and the other boundary is a line drawn in such a manner that each point on it is two hundred nautical miles from the baseline from which the territorial sea is measured.

D. *US Exclusive Fishery Management Authority (Sec. 102)*

1. All fish within the fishery conservation zone (FCZ).

2. All anadromous fish throughout migratory range of each species beyond the FCZ, except when in a foreign nation's territorial sea or FCZ to the extent such jurisdictions are recognized by the United States.

3. All continental shelf fishery resources beyond the FCZ.

4. No fishery management authority over highly migratory species

(tunas) (Sec. 103).

E. *Foreign Fishing (Sec. 201)*

1. Authorizes foreign fishing under certain conditions.

2. Provides for negotiation of Governing International Fishery Agreements (GIFAs) which:

a. Recognize US exclusive fishery jurisdiction;

b. Stipulate compliance with all regulations including boarding, arrest, seizures;

c. Require

(1) Permits be displayed;

(2) Fees be paid in advance;

(3) Agents be appointed and maintained in United States;

(4) Responsibility be assumed for loss and damage to vessels and gear.

d. Require, if U.S. requests

(1) Transponders be installed; and

(2) Observers be permitted.

3. Defines total allowable level of foreign fishing (TALFF) as that portion of optimum yield (OY) of a fishery which will not be harvested by vessels of the United States.

4. Provides for allocation TALFF. The Secretary of State in cooperation with the Secretary of Commerce shall determine the allocation among foreign nations. They shall consider whether nations

a. Have traditionally engaged in fishing in such fishery;

b. Have cooperated with United States in research and identification of fishery resources;

c. Have cooperated with United States in enforcement with respect to conservation and management; and

d. Have responded to other matters as appropriate.

5. Requires reciprocity—same treatment for US vessels.

6. Provides for preparation of preliminary fishery management plan (PMP), which is in effect until a fishery management plan (FMP) is prepared and implemented.

a. Must include preliminary description of the fishery including OY, capacity and extent to which US fish processors will process the US share of OY, and total allowable level of foreign fishing with respect to such fishery;

b. Must require permit;

c. Must require submission of pertinent data; and

d. Must contain conservation and management measures applicable to foreign fishing.

F. *International Fishery Agreements (Sec. 202).* Calls for Secretary of State to negotiate treaties:

1. GIFAs;

2. Renegotiation of old treaties;

3. Boundary negotiations.

G. *Congressional Oversight of Governing International Fishery Agreements (Sec.203).* Establishes procedures for Congressional oversight.

H. *Permits for Foreign Fishing (Sec. 204)*

1. Provides that no foreign fishing vessels shall engage in fishing within the US fishery conservation zone or for anadromous or continental shelf fishery resources beyond such zone;

2. Establishes procedures for contents and processing of permits to Councils, Secretary of Commerce, US Coast Guard, and US Congress;

3. Authorizes establishment of conditions and restrictions;

4. Establishes procedures for disapproval of permits;

5. Provides that fees shall be paid to the Secretary of Commerce by owner or operator of foreign fishing vessel for which a permit is issued. Fees shall apply without discrimination to each foreign nation.

6. Provides for registration permits.

I. *Import Prohibitions (Sec. 205)*

1. Establishes criteria and procedures for prohibition of imports of certain fish and fish products into the United States.

2. Establishes procedures for removal of prohibition.

J. *National Standards for Fishery Conservation and Management (Sec.301).*
Establishes national standards with which fishery plans shall be consistent:

1. Conservation and management measures must prevent overfishing while achieving the optimum yield from each fishery, and

2. Be based on best scientific information available.

3. An individual stock of fish shall be managed as a unit throughout its range, and interrelated stocks of fish shall be managed as a unit or in close coordination (to the extent practicable).

4. Measures shall not discriminate between residents of different states. Necessary allocation or assignment of fishing privileges among various US fishermen shall be
   a. Fair and equitable to all such fishermen;
   b. Reasonably calculated to promote conservation; and
   c. Carried out in such manner that no particular individual, corporation, or other entity acquires an excessive share of such privileges.

5. Measures shall promote efficiency in the utilization of fishery resources (where practicable), except that no such measure shall have economic allocation as its sole purpose.

6. Measures shall take into account and allow for variations among, and contingencies in, fisheries, fishery resources, and catches.

7. Measures shall minimize costs and avoid unnecessary duplication (where practicable).

K. *Regional Fishery Management Councils (Sec. 302)*

1. Establishes eight Regional Fishery Management Councils, as

follows:

a. New England Council: 17 voting members
b. Mid-Atlantic Council: 19 voting members
c. South Atlantic Council: 13 voting members
d. Caribbean Council: 7 voting members
e. Gulf of Mexico Council: 17 voting members
f. Pacific Council: 13 voting members
g. North Pacific Council: 11 voting members
h. Western Pacific Council: 11 voting members

2. Establishes rules of procedure and administrative guidelines for Council operation; and

3. Establishes Committees and Panels.

L. *Contents of Fishery Management Plans (Sec. 303)*

1. Any FMP prepared by Council or Secretary shall:

a. Contain conservation and management measures which are necessary and appropriate for the fishery, and consistent with the national standards;

b. Contain a description of the fishery;

c. Assess and specify the conditions of the stock, the maximum sustainable yield (MSY) and OY.

d. Assess and specify annual US domestic harvesting capacity; US processing capacity; the portion of OY which, on an annual basis, will not be harvested by US fishery vessels and can be made available for foreign fishing; and the capacity and extent to which US fish processors will process the US share of OY, on an annual basis.

e. Specify pertinent data which shall be submitted to the Secretary of Commerce.

2. Any FMP prepared by Council or Secretary *may:*

a. Require a permit;

b. Designate zones and/or seasons and/or gear;

c. Establish specific limitations on catch of fish;

d. Prohibit, limit, condition, or require use of gear, vessels, or other equipment to facilitate enforcement;

e. Incorporate (consistent with the Act) relevant conservation measures of the coastal states nearest to the fishery;

f. Establish a system for limiting access; and

g. Prescribe other measures as are determined to be necessary and appropriate for conservation and management.

M. *Action by the Secretary (Sec. 304)*

1. Prescribes action required by the Secretary for review, approval, partial disapproval, and implementation of plans and amendments.

2. Prescribes conditions under which Secretary may prepare a fishery management plan and implement such a plan.

3. Provides for collection of fees, which shall be limited to administrative costs incurred in issuing US permits.

4. Requires the Secretary to conduct fishery research.

N. *Implementation of Fishery Management Plans (Sec. 305).* Provides procedures for plan implementation through promulgation of regulations.

O. *State Jurisdiction (Sec. 306).* Provides procedures for regulation of fishery under an FMP within state waters under certain limited conditions.

P. *Prohibited Acts (Sec. 307).* Lists prohibitions under the Act.

Q. *Civil Penalties (Sec. 308).* Establishes civil penalties and procedures for assessment and review.

R. *Criminal Offenses (Sec. 309).* Specifies offenses which carry criminal penalties, and establishes penalties thereto.

S. *Civil Forfeitures (Sec. 310).* Establishes authority and procedures for forfeitures.

T. *Enforcement (Sec. 311).* Establishes responsibility and powers of Secretary of Commerce and US Coast Guard to carry out enforcement of regulations.

U. *Miscellaneous (Sec. 312, 401-406)*

1. Effective date of certain provisions of the Act;

2. Effect of Law of the Sea Treaty;

3. Repeal of other Acts;

4. Fishermen's Protective Act Amendments;

5. Marine Mammal Act Amendment;

6. Atlantic Tunas Convention Act Amendment; and

7. Authorization of Appropriations.

II. Other Applicable Laws

There are a number of other federal laws that affect the Councils and their operations. One category of laws interacting with the FCMA involves those that affect the federal decision process through prescribing certain environmental or economic considerations (NEPA, E.O. 12044), or through establishing rules of procedure for public participation or access (APA, FACA, FOIA). Another category includes those that deal with the competing uses of the ocean, the preservation of ecological systems, and the increasing demands on the nation's coastal areas.

A. The Decision Process

1. *National Environmental Policy Act (NEPA)*

The National Environmental Policy Act expressed the intent of the Congress to achieve coordination of federal activities with environmental considerations. The basic purpose of NEPA is to insure that federal officials weigh and give appropriate consideration to unquantified environmental values in policy formulation, decision-making, and administrative actions, and that the public is informed at an early stage, and is provided adequate opportunity to review and comment on the major federal actions. NEPA requires preparation of a detailed Environmental Impact Statement (EIS) for major federal actions that significantly affect the quality of the human environment. The major federal action in relation to the FCMA and FMP development has been determined to be the promulgation of final regulations.

Environmental Impact Statements will be prepared for all fishery management plans as well as significant amendments to existing plans. In the case of amendments, a determination of environmental significance must be made through the preparation of an environmental assessment, which may or may not result in the preparation of a Supplemental EIS. Specific NEPA regulations on procedures and format were issued by the Council on Environmental Quality on November 29, 1978.

2. *Executive Order 12044*

Executive Order 12044, issued in March 1978, outlines the President's policy for developing regulations; sets minimum procedures for developing significant regulations, including publication by agencies of a semi-annual agenda of regulations under development or review; requires close agency head oversight; calls for early and meaningful opportunities for public participation; requires careful review of the need, quality, and effectiveness of regulations being proposed; requires careful economic analysis of regulations having a major economic effect; and requires periodic review of existing regulations. Councils are affected by NOAA's adaptation of the Department of Commerce regulations implementing the Executive Order.

3. *Federal Advisory Committee Act (FACA)*

The intent of Congress with regard to the application of the Federal Advisory Committee Act is stated in the Report of the Conference Committee accompanying the FCMA. The provisions of FACA apply to the Councils and their Committees and Panels. Application is designed to ensure open meetings and public access to information generated by the Councils.

4. *Freedom of Information Act (FOIA)*

The Freedom of Information Act provides for public access to records of the executive branch of the federal government and to records generated at the request of the federal government. Nine groups of exceptions are provided that allow the withholding of information. FOIA requests made to the Councils for release of information are handled by the Assistant Administrator for Fisheries who makes his decision in consultation with the Councils receiving such requests.

B. Uses and Ecology of Ocean and Coastline

1. *Marine Mammal Protection Act (MMPA)*

Although marine mammals are not covered by the FCMA since "fish" is defined in section 3 (6) as excluding marine mammals, section 404 of the FCMA amends the MMPA to extend the waters to which that Act applies to two hundred miles from the coastal baseline. The moratorium on taking of marine mammals thus became effective in the FCZ on March 1, 1977. Under section 101 (a) (2) of the MMPA, permits for the taking of marine mammals incidental to the course of commercial fishing may be issued subject to regulation. With passage of the FCMA, this section of the MMPA applies to foreign as well as to domestic fishing in the FCZ.

2. *The Fishermen's Protective Act (FPA)*

The FCMA amends the Fishermen's Protective Act generally to the effect that the federal government will reimburse the operator of an American fishing vessel under certain conditions for fines and losses resulting from a seizure of that vessel by a foreign nation for fishing within waters adjacent to that nation. A new section 10 of the FPA also provides for compensation to US fishermen for fishing-vessel damage or loss caused by foreign vessels, and fishing-gear damage or loss caused by foreign or domestic vessels or acts of God in the FCZ.

3. *Coastal Zone Management Act (CZMA)*

The principal objective of the CZMA is to encourage and assist states in developing coastal zone management programs, to coordinate state activities, and to safeguard the regional and national interests in the coastal zone. In the preparation of fishery management plans, Councils should be particularly cognizant of section 307(c) of this Act, which requires that any federal activity directly affecting the coastal zone of a state be consistent with that state's approved coastal zone management program, since activities taking place beyond the territorial sea may impact in the coastal zone. Council FMP development may also be indirectly aided by the initiation of a Coastal Fisheries Element (CFE) as part of the CZMA program development and implementation grants. The purpose of CFE is to provide the states with financial and technical assistance to develop information required for more effective management of fisheries within the territorial sea.

4. *Endangered Species Act (ESA)*

This Act provides for the conservation of endangered and threatened species of fish, wildlife, and plants. The program is administered jointly by the Secretaries of Commerce and Interior. Councils should be mindful of the threatened and endangered species list; when preparing FMPs, they should consult the National Marine Fisheries Service and the US Fish and Wildlife Service, as a routine procedure, as to whether the plan jeopardizes the continued existence of a listed species or results in the destruction or modification of its critical habitat.

5. *Marine Protection, Research and Sanctuaries Act (MPRSA)*

The MPRSA authorizes the Secretary of Commerce, after consultation with appropriate federal agencies and with Presidential approval, to designate ocean waters as far seaward as the outer edge of the continental shelf as marine sanctuaries, to preserve or restore distinctive conservation, recreational, ecological, or esthetic values. Since passage of the FCMA, NOAA* has established certain policies and procedures regarding sanctuary designation (embodied in proposed regulations) which ensure Council participation and coordination with fishery management plans.

PROBLEMS OF IMPLEMENTATION

Eight FMPs are in final stages of approval; 31 FMPs or major amendments are scheduled for submission by the Councils during the remainder of 1979. Each of these FMPs has presented new legal and policy issues, which have brought about delays in implementation because they have needed the careful analysis commensurate to precedent-setting decisions. But we learned from this and are prepared to face subsequent problems with greater confidence in the Act.

Generally, problems associated with implementation of the FCMA are now recognized. The problem areas serve as examples of a national failure to allocate resources efficiently and/or communicate with the affected public. It is important that we understand the possible implication of this failure for the future, and, more importantly, that we develop means to deal with it effectively.

In fishery management, as in most regulatory arenas, crises are matters of degree, being emotionally linked to such terms as "resource emergency," "economic chaos," and "civil disobedience." It is not necessary to define crises in order to discuss problems generally common to their management, including the scarcity of accurate information—for example, catch/effort, status of stocks, cost/earnings, economic dependency, and the changing character of the players as the negotiations (within the Councils or at the agency level) leave one or more parties dissatisfied.

Perhaps the greatest frustration is the uncertainty concerning what kinds of management regimes will develop, and how they will affect the lives of fishermen. Several of the problems are not generic. They result from difficulties associated with specific fisheries, and, once resolved in those fisheries, should not affect the overall management process. However, these problems have tasked the Councils and the staff of the NMFS considerably, and have been very difficult to resolve.

A. *The FMPs*

For example, The FMP and amendments developed by the Pacific Fishery Management Council for management of the commercial and

*The National Oceanic and Atmospheric Administration (NOAA)

recreational salmon fisheries off the coasts of Washington, Oregon, and California have been very controversial. The Council was charged with developing a plan that would allocate highly prized but limited West Coast salmon resources among commercial troll fishermen, charter boat operators, inside recreational fishermen, inside commercial fishermen, and treaty and non-treaty Indians at the same time that it permitted an adequate escapement of salmon to spawning grounds. The management measures that were adopted, the Council felt, were the best compromise it could devise. The Secretary approved the 1979 amendment in April and NOAA has since faced complaints and legal action.

The Mid-Atlantic Fishery Management Council's FMP for the butterfish fishery of the Northwest Atlantic was disapproved this spring because the Council used inadequate information in specifying optimum yield. The FMP that the Council submitted established an OY 5,000 metric tons below a maximum sustainable yield for the fishery of 16,000 metric tons. This was done in order to effect a corresponding reduction of total allowable level of foreign fishing of 5,000 metric tons. The Council believed that this reduction of foreign fishing would improve export market opportunities for the developing domestic fishery. The FMP was disapproved since it failed to provide a summary of supporting information on which the reduction was based, as required by section 303(a) (3) of the FCMA. A recommendation was made by NMFS as to how the deficiency could be corrected. The staff of NMFS subsequently worked with the Council staff on providing proposed changes for the Council's consideration.

Managing the severely overfished New England groundfish fishery for cod, haddock, and yellowtail flounder has been an extremely difficult task. Despite repeated closures and reductions in trip limits, the FMP conservation measures were unable to stabilize the fishery because of: (1) the difficulty of managing multi-species fisheries; (2) the substantial influx of additional vessels and increases in fishing capacity; (3) lack of agreement in the Council about the objectives of the FMP; (4) lack of acceptance by fishermen of a management regime that limits their fishing; and (5) initially, a lack of state matching management measures. The New England and Mid-Atlantic Fishery Management Councils—in conjunction with the National Marine Fisheries Service, the states, and other federal agencies—are working toward development of a comprehensive FMP. This process will evaluate other options, strive for better public understanding, and, in time, should provide assistance to the Management Councils in their task.

B. *Boundary Issue*

The FCMA describes generally the territorial jurisdictions of the Fishery Management Councils, leaving it to the Department of Commerce to establish exact boundaries between adjacent Councils. Boundaries had been generally agreed to by all but the South Atlantic/Gulf Councils, which wanted an exact demarcation line established—each at a different location. Interim regulations announcing the Secretary's preliminary decision on the boundary lines were published in July 1977. The South Atlantic/Gulf boundary, which was set at the Dade-Monroe County line in Florida, was supported by the Gulf of Mexico Fishery Management Council, since it placed several important Florida key fisheries under the Gulf Council's management. But the South Atlantic Fishery Management Council objected, and urged instead a more southwesterly boundary line. NOAA requested that the Councils resolve the issue but they were unable to do so. An initial agreement to submit the dispute to binding arbitration fell apart in September 1978. Each of the parties then submitted memoranda and requested the Secretary to resolve the dispute. On March 2, 1979, the original boundary was reaffirmed. The South Atlantic Council, in response, decided to seek a judicial review of the decision, but there is doubt that the Council has standing to sue.

The South Atlantic/Gulf boundary dispute is unique and, like the problems of salmon of the Pacific Northwest and groundfish in the Northwest Atlantic, is being addressed with a view to its very individual nature.

C. *Communications*

We have also been faced with problems, however, that are more universal.

First and foremost has been the reaction by domestic fishermen to the FCMA itself. The Act clearly provided for conservation and rebuilding of fish resources and requires management of both domestic and foreign fishing effort. But many fishermen did not understand that the management provisions of the Act applied to them as well as to foreign fishermen. There have been—and will continue to be—situations in which domestic fishing must be tightly managed to stabilize or rebuild stocks.

Although the FCMA has begun to reverse the trend in stock depletion and has reduced foreign fishing competition, some of our domestic fishermen have suffered reduced catches, frustration, and economic loss due to management measures restricting their traditional fishing operations. In some cases, increased domestic catches have resulted in lower prices to the fishermen but, rarely, to the consumer. In many of our major fisheries, too many fishermen want to share in a limited supply of fish. One answer to this problem is to expand United States efforts for nontraditional underutilized fisheries. Development of these fisheries is, in fact, one of the goals of the Act.

D. *The OY Problem*

Despite a great deal of study of the subject, including a technical workshop and a national conference, there still appears to be considerable confusion as to the meaning and operation of the optimum yield concept. We are formulating guidelines for the determination of OY that will provide for maximum flexibility. The question also is being addressed by the Pacific Council, whose Scientific and Statistical Committee has drafted a statement. I believe that, as a result of our joint effort with the Councils, useful guidelines can be evolved. A major difficulty is the need to be definitive enough to be useful, without being unduly restrictive.

One point of confusion appears to be that many people regard OY as a number or as a quota. In our view, this is too narrow. It is rather a condition or set of conditions for a fishery that "will provide the greatest overall benefit to the nation, with particular reference to food production and recreational opportunities." For example, in the FMP for Pacific anchovy, OY is a formula which determines the annual bait and reduction fishery quotas according to the stock abundance predicted for the year. In the case of Stone crabs, by comparison, the OY is all those harvestable male crabs with a claw length greater than ten centimeters. In the surf clam and ocean quahog FMP, the OY is expressed as a number.

This flexible approach does not eliminate the necessity for an annual estimate of weight or number of fish representing the OY for the year. The estimate is needed to determine the total allowable level of foreign fishing required by section 303(a)(4)(B) of the FCMA. But the statement of criteria defining optimum yield may remain the same from year to year, while the number may change according to the circumstances of the fishery. The major advantage of specifying OY as a set of criteria in an FMP, where this is possible, is that it can avoid the need to go through the lengthy amendment process each year. I am optimistic that use of this flexible concept will greatly simplify the implementation of the Act.

E. *Other FCMA Problems*

1. Domestic Annual Harvest

The concept of domestic harvest (DAH) is also critical in developing FMPs. The DAH is the inclusion of an assessment of the capacity and extent to which fishing vessels of the United States on an annual basis will harvest the optimum yield. This amount is set aside to give first priority to domestic fishermen. The surplus is available for foreign fishermen.

The estimation of this figure is a difficult matter. National Standard Number One of the FCMA requires achievement of the optimum yield. We take this to mean that the management measures should permit the combined domestic and foreign catch to reach the OY determined for the fishing year. Under-use would be wasteful of the resource. Over-use could deplete the resource.

If a Council sets the estimate of domestic harvest too high, the surplus available for foreign fishing will be correspondingly lower and the total catch will fall short of OY. If, in an attempt to avoid this problem, a Council sets OY too low, it may be necessary to reduce foreign allocations in mid-season. This is clearly an undesirable action.

One way to avoid the problem is to put a part of the OY into a reserve account. This can be allocated to domestic or foreign fishing as catches show it is needed. Such a procedure requires an efficient mechanism to monitor fishing, predict trends, and to reallocate reserves in time for them to be harvested.

Some FMPs have not identified a reserve as such, but have inflated the estimate of US harvest to provide the necessary protection and have provided for reallocation from this account to TALFF when fishing results show that US harvest for the year is likely to fall significantly short of the estimate. We are, in essence, experimenting with different mechanisms to manage fisheries that may require in-season changes of harvest estimates. Either mechanism may prove acceptable, but it is not reasonable to use both an inflated estimate of US capacity and a substantial reserve.

2. The Joint Venture Situation

The curtailment of foreign fishing effort has led to considerably greater foreign interest in investment in US fisheries and/or in joint venture operations. Joint US/foreign fishing ventures are permitted, after application, unless the Secretary determines, as directed by the FCMA under Sec. 204(b)(6)(B), that

> ...United States fish processors have adequate capacity, and will utilize such capacity, to process all United States harvested fish from the fishery concerned.

This section also limits the amount of fish that may be received under a joint venture permit to

> ...that portion of the optimum yield of a fishery concerned which will not be utilized by United States fish processors.

NOAA feels that the issue of foreign investment and joint ventures is significant and is attempting to document the extent to which foreign participation in the US fishing industry is occurring and to analyze what impact this is having or may have on the interests of US fishermen and processors.

3. The Interjurisdiction Problem

Under the FCMA, the Secretary of Commerce has authority to enforce regulations only for stocks of fish harvested outside three miles. Unless the fishery is predominantly within the fishery conservation zone (FCZ), the Secretary cannot implement a Council-approved plan within the territorial sea because the Act left essentially unchanged the authority of the coastal states to regulate fisheries within the territorial sea. Inland waters, such as Chesapeake Bay and Puget Sound, are not covered by the Act.

This issue is especially significant because the resources involved include some of our most valuable commercial and recreational fisheries; at least fifty percent of the domestic commercial harvest and approximately eighty percent of the recreational catch is involved. Examples are menhaden (the largest volume fishery); striped bass (a major recreational species found off every coastal state from Maine to Washington state, and a commercial fish in some states); and shrimp in the Gulf of Mexico and South Atlantic (our most valuable fishery). Lack of a uniform management system for interstate fisheries has caused user conflicts and resource depletion, and created inequities and economic hardships to both commercial and recreational fishermen.

NOAA has given the matter of effective territorial sea management a very high priority, and is in the process of assessing its resources and legislative authorities to develop a strategy for pooling them to ensure most effective management.

4. The Enforcement Problem

Domestic enforcement problems have arisen in connection with several of the FMPs. Fishermen have attempted to avoid restrictions on fishing imposed by FMPs by claiming to have harvested a species within the state three-mile zone rather than in the FCZ. There is strong evidence, for example, that fishermen taking groundfish in New England are using this tactic to harvest cod and haddock, far in excess of amounts allowed by regulation.

Most states, however, either have adopted (or are in the process of adopting) regulations compatible with the federal regulations. Although not all state regulations are parallel with FCMA regulations, they do provide an improved basis for more effective regulation of fishing.

Obviously, the modern crises of fisheries management present decision-makers with agonizing management choices. Very often the manager is confronted with a plethora of conflicting information and is given very little time to choose an appropriate course of action. Although contemporary methods of systems analysis have been used in attempts to organize data and clarify options, a lack of data, misunderstanding, poor communications, or some combination thereof results in such methods being of little use in presenting an accurate picture of the various individual values and perceptions. Thus, it is clear we must promote greater cooperation and understanding among scientists, resource managers, fishermen, and others. The ultimate goal is a system by which individuals of all persuasions cooperate in fisheries management on the basis of mutual benefits.

CONCEPTUAL ASPECTS OF RESOURCE MANAGEMENT

Because of the technological force of fishing and the depletion or collapse of many highly valued fisheries, many nations of the world have extended their national jurisdiction to encircle most of the more productive coastal ecosystems and have begun intensive management efforts. Most management of fishery resources, however, has used a single species approach, and generally within a limited geographic range or environment. This encircling movement has simultaneously created an awareness of the need to address other aspects of fishery management, environmental considerations, and, to a lesser degree, fishery development.

Although it is not my intent to dwell extensively on any given aspect, I do wish to explore some general proposals. I would not be surprised to hear that others internationally have identified similar aspects and are dealing with them in an aggressive manner.

A. *Total Fish Ecosystem*

Within the last few years multi-species management has received some attention, yet those who have proposed such an approach have had little impact upon the strategy of management employed by the fishery manager. Presently several Fishery Management Councils are struggling with multi-species modeling as an assist to the manager. Their approach arises from the perspective that we must reduce the number of differently regulated fisheries in order to deal effectively with a single unit. There appears to be a growing recognition that, in fishing, multiple regulations may be perverse enough in combination to drive the total fishery ecosystem to an undesirable state. I believe this recognition is slow in coming and the issue must be dealt with more rapidly.

A total ecosystem approach was first made at the Annual Meeting of the International Commission for the Northwest Atlantic Fisheries (ICNAF) in June 1972. The two-part ICNAF Document 72/119, entitled "A Preliminary Evaluation of the Effects of Fishing on the Total Fish Biomass and First Approximation of Maximum Sustainable Yield for Finfishes in ICNAF Division 5Z and Subarea 6," described the changes in relative biomass of groundfish (Part one) and the relationship of total fishing effort to the yield (Part two).

This total ecosystem approach recognized that the species sought in the area between Cape Hatteras to Nova Scotia were many and that the various fisheries significantly interacted. The approach described in the document was developed for many reasons, two of which are most important to the fishery manager and user groups: (1) the by-catch or incidental catch problem; and (2) the desire to maintain the ecosystem in its most productive state. The first of these has been reasonably well documented and can be stated simply: few, if any, fisheries (and certainly no trawl fishery in this area) can be classified only by harvest of the prime-target species. Inasmuch as the mixed-species situation is pronounced in many continental-shelf areas throughout the world, priority should be given to evaluation of management and fishery-development alternatives. This type of regulation requires a knowledge that is quite advanced.

B. *Fishery Development*

The Fishery Conservation and Management Act of 1976 was passed to conserve and manage fishery resources off the coasts of the United States. Two major objectives of this program are (1) to promote domestic commercial and recreational fishing under sound conservation and management principles, and (2) to encourage development of fisheries which are currently underutilized or not utilized by United States fishermen. In implementing the legislation, the primary emphasis has been to rebuild depleted stocks and maintain other stocks at an optimum level for exploitation.

The implicit relationship of management to fishery development has not received the attention it merits, in part because the first actions called for by the Act required the establishment of Regional Councils and the preparation of management plans. However, development is affected by management strategy, which in turn should be responsive to development opportunities. Fishery managers should be aware of this reciprocal relation; they should be constantly attuned to changes in the fishery to ensure that the industry can progress efficiently. The responsible institutions (state and federal as well as the Councils) must have an overview (or at least be appreciative) of both management and development in order to cope with social and economic problems—such as growing demands for recreational as well as commercial opportunities, conflicts among user groups, and allocation of the resource.

The Act specifically encourages fishery development. In this regard, the most important aspect of the FCMA, undoubtedly, is our authority to limit foreign fishing to that portion of the OY which will not be harvested by US vessels. Not only has this stipulation eliminated excessive exploitation of certain species by foreign vessels, but it has led to increased US production of these species. Rehabilitation of depleted stocks—such as haddock and cod in the Atlantic, and ocean perch and sablefish in the Pacific—through sound management practices will directly contribute to the strengthening and development of US fisheries. Management of species not fully utilized by the domestic fleets, such as butterfish and Alaska pollock, also will contribute to expansion of the industry. By maintaining these stocks at levels consistent with MSY, domestic interests will be afforded the opportunity of increasing their share of the OY in accordance with their capacity to harvest such stocks.

The FCMA marks a new era in US fisheries. As the Council and the state and federal governments gain experience through the development and implementation of FMPs, changes in procedures, and perhaps in the legislation itself, will enhance the effectiveness of the Act.

Fishery management programs should be designed to increase the domestic harvest opportunity for unused or underutilized fishery resources and be in harmony with development programs designed to increase industry's capabilities. Stock assessment, along with information acquired from the users, provides a basis for deciding which resources offer potential for commercial or recreational development. Exploratory fishing programs that involve gear development, testing, and demonstration can provide specific information on target species. Opportunities to apply new techniques and new fishing strategies for more efficient harvesting can be identified for fishery development.

Another FCMA avenue of furthering development is that permits can be issued to foreign vessels to receive catches from US fishing vessels for foreign processing or foreign markets. These "joint ventures" offer several potential advantages for the development of the domestic industry. In addition to the obvious benefit of creating new markets, the US can gain by obtaining more data on underutilized species. Further, if the catches are concentrated on predatory species (pollock for example), the abundance of more valuable species, such as salmon, may increase. "Joint ventures" may provide opportunities for US harvestors to benefit from foreign fishing technology as well. "Joint ventures" are not without disadvantages, however, and the potential "competition" is objectionable to elements of the US processing industry. There also is concern that the incidental catch in such operations may lead to excessive harvests of higher valued species, and that "joint ventures" provide another avenue for foreign access to US fishery resources.

Fishery managers, especially those associated with Fishery Management Councils and state fishery development agents, should consider fishery development aspects and their social implications—so that development is not impeded and so that the recreational needs of the nation are satisfied.

C. *Environmental Considerations*

Maintenance of water quality and natural habitat is a major long-range problem in the management of many fish stocks. For example, the cumulative destruction of wetlands from development in the

Gulf of Mexico estuaries adversely affect spawning, nursery, and rearing areas. Marine species, including menhaden, flounders, weak fish, and croakers, are dependent on estuaries early in their life cycles. Also, anadromous fish stocks depend heavily on the quality of waters in which they spawn. This is recognized in the FCMA, which makes special references to the importance of environmental protection to the health and abundance of fish stocks.

The term "conservation and management," which occurs frequently in the Act, is defined to include the environment. Section 304(e) instructs the Secretary to initiate and maintain a biological research program into the impacts of pollution on fish and the impacts of wetland and estuary degradation, among other things. Reflecting these provisions of the law, section 602.3(b)(6) of the Council Regulations requires that each fishery management plan should: (1) describe habitat of the stock and factors affecting its productivity; (2) identify habitat areas of particular concern; and (3) describe programs of habitat protection or restoration.

Thus, Councils may make habitat protection recommendations in FMPs, and we encourage them to do this. Upon approving such plans, the Secretary implements those recommendations for which he has statutory authority and adequate resources. For example, through its Fish and Wildlife Coordination Act programs, NOAA participates in decisions on the proposed refineries, powerplants, real estate development, and industrial effluents. Council recommendations can be coordinated into State Coastal Zone Management Programs. Programs of research can be undertaken as well under section 304(e) into man-caused factors affecting habitat productivity and identification of areas of particular concern. Remaining recommendations are directed to the agencies that have appropriate authority, such as the concerned states and the Environmental Protection Agency.

Models developed for management of single species have not taken into account the effects of the environment. I suspect that there may be factors within the environment that have effects equal to or greater than the effects of fishing. We must deal with these aspects and treat the "right disease."

CHALLENGE FOR THE FUTURE

We have discussed some of the many challenges to the present day fishery resources manager. The natural marine ecosystem represents the optimum in terms of balance and survival of its living resources. The major challenge is, thus, whether or not man can retain a balance and maintain the yield that the world expects.

REFERENCES

[1] Smith, Roland F. "Management of Living Resources," paper prepared for Third Annual Seminar of Center for Oceans Law and Policy, January 1979.

[2] U.S. Department of Commerce, "U.S. Ocean Policy in the 1970's," October 1978.

Commentaries

KEN CAMPBELL
Fisheries Council of Canada
Canada

At the present time we are in a state of formation of policy again. We have had this kind of situation many times in the past. The last time we tried to put fisheries policy down on paper was in 1976. This was triggered by the crisis of 1974, which had really brought the fishing industry—in particular the Atlantic ground fishing industry—very close to collapse. The government at that time decided to look very carefully at the industry, how it operated and its structures.

Interestingly enough, the lead policy stated at that time was that we should extend jurisdiction. Of course, the decision had already been made before that, but this was or seemed to be the key point from which other policy objectives and rehabilitation of the industry have come. That has happened, and we are as a result in a new game in fisheries and management fisheries exploration. It gives us an opportunity to improve the stability and viability of the industry. We can look forward in Canada to some pretty impressive increases in fish production.

Increases for example on ground fish—particularly cod—at least double and possibly a factor of three over 1977 levels by 1985. The salmon program on the West Coast also has an objective of doubling that resource from the present level. When we took over the resource we knew we had a depletion problem, but we didn't know how great it was.

The first policy was to set catch limits to permit rehabilitation of these depleted stocks, and it was hoped that this could be done with as little as possible disruption or impact on Canadian industry. That meant it had to be done largely at the expense of the foreign fleet. That idea didn't find favor in Canada because we became the foreign fleet for causing the problem in the first place.

It had to happen also that the punishment of the rehabilitation period had to be spread out among various Canadian fishery sectors as well. Now there is what I would call a policy gap: we need a greater commitment in Canada to stock research. We are faced with trying to keep government expenditures down, to reduce them if possible, and unfortunately governments don't take a favored look at fisheries or fisheries research when they start cost-cutting programs. So instead of more resources—financial and professional—being made available for this job, we find that they are being cut back.

The policy was also to try to gain effective management regime outside the 200-mile limit. This meant gaining the cooperation of foreigners by allocating catches inside the 200-mile limit, and also allocating some quotas not considered surplus.

On that point I take issue with Parzival Copes' chapter. These two main policies have been a very taxing exercise for government and industry in the past two-and-a-half years, especially when just recovering from a serious crisis which began in 1974. On top of that we have had a Ministry of Fishery which is interested in some basic restructuring. The Federal Minister raised questions about the integration of the fishing industry and its rationalization. He questioned favoring in-shore fisheries at the expense of the off-shore, rather than establishing a policy which we think would be more sensible—achieving a proper balance between the two.

The Federal Minister doesn't have jurisdiction over the secondary sector, but he has "absolute authority"—those are the words of the Federal Fishery Act—in the licensing of fishing. Of course, this has to have its effects on the secondary sector. So these problems have created a policy vacuum which has to some extent inhibited the development of objectives as to what kind of fleet we should have to exploit the stocks that we now have—which we had not utilized in the past, or from rehabilitated stocks. And also, objectives as to what kind of processing facility is needed to meet landings, and finally even objectives in the development of markets. That, of course, has affected decisions on mobilization of capital for our industry, at least for tremendous amounts of capital which will be needed for building new vessels.

We have an additional policy. It's a short-term one—so we are told and hope to God it is. Our fishermen sell over the side to foreign vessels. This started out to provide a counter or a market for fish which the fishermen happened to catch but couldn't sell to Canadian processors because they didn't have the capacity to handle them or the markets to move them. However, those things have changed. Yet the policy is continuing, and we look to this side of the border, where your government has taken a conscious decision that the processing sector should have at least first call on what is caught by American fishermen. This too has had a negative impact on our own ability to develop a means to catch, process and market our resources.

We are not permitted by the federal government to own or operate freezer trawlers, and yet we have very valuable resources, like squid, in our own zone, which require freezing on board. We have to allocate them to foreign vessels which take them home to market. We feel we can do just as good a job as they can. These last two years a serious problem has been the consultation process, especially on the Atlantic Coast. We have to improve in that respect. The US Management Councils may not be as good in practice as they look to us, but they have a couple of advantages. One we see is a pretty good cross-section of representation of the interests and sectors of the industry. Even more importantly, they bring the management of the fisheries to the coast where the action is.

This question is important to us because one of the reasons Ottawa is called on to make so many management decisions is—as Copes said—that there are five regional governments, five provincial governments and, even though the federal government has sole authority for management, there are three federal regional offices in there, and they don't always see eye to eye.

We find that their region often is in conflict with the region in martime, and Ottawa has to be the arbitrator. Then we find management decisions being taken out of our coastal area. The chapter talks about debate on the jurisdiction question. I agree that the federal government will keep the authority for management. One of the reasons the provinces are looking for some management authority is the lack of consultation by the federal government with the provinces on management questions which the provinces feel affect the well-being of their people and their communities. If we were to divide the jurisdiction by coastal provinces, however, this would be an impossible situation because the resources would be common to all of them.

There has to be better input from the provinces into the federal management decision. Provinces are objecting to the absolute authority used by the Federal Minister in the past couple of years.

The chapter talks about the proximity principle. It is the concern of the industry that this proximity principle is starting to gain some recognition by the federal government.

The point Copes makes about the tendency for the in-shore fishery to grow unchecked isn't being made loud enough. Politicians look on the in-shore fishery as the biggest bank of votes when it comes to elections and things that happen in between elections. Unfortunately they have taken the attitude that the in-shore fisheries should not be controlled. Of course, to have the balance I am advocating would permit rational development of an industry which is productive, efficient, and internationally competitive.

JACOB DYKSTRA

Point Judith Fishermen's Cooperative
Rhode Island
United States

William Gordon writes of international commissions and their ineffectiveness. I would like to add something about this. Somebody may be able to argue that there have been one or two international commissions in which the United States has been involved that have been effective, but I have argued—successfully sometimes—that none of them have been effective. They just didn't work out. Yet there are still people who will defend these commissions.

This is symptomatic of something that drives me up a wall. When people speak of some of the management schemes which are currently being described, they say: "There are a few bugs in them. There are a few problems, but we will work them out as we go along." Sometimes this is said by the very managers of these plans who have just discoursed about how they are not working. Afterwards they will say "Well, they are really not that bad." One gets this all the time: "Well, some of these schemes are not that bad." Actually, they are causing all kinds of social and economic disruption. They are not working at all, and nobody knows just what to do about it.

Gordon writes about the eco-system approach and the mixed California fishery. Most of the fish are caught by this method, and also there are extensive seining operations. The really big operations—the one we have had trouble with—have been trawl fisheries. Some of them have a single target species, but most of them have several target species, as well as all kinds of bi-catch problems.

Nowhere in the world has there so far been effective management of this type of fishery. People will defend them, but as I see it so far we have not learned how to have effective management of this type of fishery, and that's the problem that has been bugging us here on the New England council.

I would like to describe how this law came into effect, how it was created, and how these things happened in the United States. First of all, executives of the federal government fought us bitterly on this law. They didn't want it and they tried to divert it, even though there are a number of them now who think it's the best thing and even claim some credit for it. But when the law was in the process of being put together it was done by staff people on committees in Congress.

Normally when an administration is behind something like this they will write a bill and send it up to the hill and staffs will start to work with that. There will be hearings. In this case there was no administration bill. Some of these staffs started right from scratch. Two people particularly—one staff member from the Senate and one staff member in the House—did most of the writing. I have in my files pages and pages and pages of crossed-out management measures that were proposed and shot down before they even got to the hearings.

It wasn't an easy process. Some of us were quite amazed at what survived and what didn't survive. Now and again we look at the thing as some sort of a master creation. Yet I know the people who burned the midnight oil to write some of these phrases would be just amazed at the way everybody hangs on them now.

Another thing that happened when this was put together was that there were some people who were much in favor of very strict domestic management. If you are in favor of something, you try to get the Senator or the Congressman to ask a question which will bring this out. One particular staffer—either he or his boss—asked whatever state official, government official, academician or whomever came along, "Do you think there should be domestic management?" He then put together this list of all those who wanted domestic management and he started writing again. That is the way we got a lot of domestic measures. A lot of the problems we have had with fisheries management plans being considered inadequate by the secretary have not been due to the provisions of the 200-mile bill itself. But there is a little phrase in there, "other applicable laws," and Gordon's chapter lists several pages of these. One of the so-called other applicable laws concerns marine mammals. In spite of everybody, in spite of environmentalists and little old ladies in tennis shoes or whom ever we are going to have to come to the point where we consider marine mammals a part of the system.

We are managing human beings, we are managing all kinds of other mammals, we are managing fish. Somebody says, "Let's not manage marine mammals, let's preserve them. Let's not harvest any of those, we will leave them there." Marine mammals are now beginning to harvest several times as much fish as human beings are, and according to Gordon we are talking about four million tons.

Let's move on to the ground fish plan of New England. How did we get in the mess we got in with the ground fish plan? We have had all kinds of controversy, difficulty, lots of changes. Most people I think have pretty much forgotten how we began. We sat down as a council, all of us very green, and said, "Well, those species have got to be managed right away. We have got to do something."

We tried to do a quick and dirty job and that is what happened. We didn't do a very good job. We were inexperienced and just ran into a lot of unforeseen problems. It is the responsibility of the council to put together the plan, but it is the responsibility of the Commerce Department to direct the regulations.

I don't know how common is the experience, but when other agencies write regulations for laws, one can't recognize any connection between the two.

Gordon claims in his chapter that Optimum Yield (OY) has to be redefined. We still insist on making OY something which is annual. We have to get to a point where we say OY is long-term yield expressed some other way, perhaps in number, but a long-term yield expression rather than an annual expression.

In fisheries management we must take a look at the situation and try to do those things that are really economical, that we can really justify. I don't think this has been done so far. What we are doing is buying time until we find something else that will work. The time for this sort of fisheries management is running out. Can we manage fisheries so that the fisherman retains his traditional way of life and at the same time make economic sense?

I hope this will not mean complete and total management of fisheries. Most managers will freely admit that the cost of managing fisheries in a very tight way is more than the fisheries are worth.

BRIAN ROTHSCHILD

National Oceanic and Atmospheric Administration
United States

I would like to address myself very briefly to the European Economic Community (EEC) discussion: the nature of the problem, across-the-board problems, elements for comparison among the different presentations, some areas of difficulty in interpretation, and Third World fishery development.

First of all, John Farnell did an excellent job summarizing the problems of the EEC for the 1970s. He noted depleted internal fishery resources, the traditional lack of self-sufficiency in the fisheries products, the contraction of the traditional fishery and non-EEC waters, the high level of foreign fishing, the over capacity of the EEC fleets, and our economic dependency upon the fisheries. These were not only the problems of the 1970s in the EEC area; they were the problems in the same area in the 1940s, the 1950s and the 1960s. The critical issue is whether this set of problems will remain all through the 1980s. We can come to some determination of this if we look at across-the-board problems and see how they are being addressed.

The first problem is sovereignty. In Canada we talked about the federal-provincial system, and the fact that this system does not admit the management of some fisheries stocks. We have the same problem in the United States with the federal system. In addition to that kind of sovereignty there is the sovereignty between the internal states in any one of these arrangements and the external ones. I needn't go into the present US and Canada considerations. John Farnell didn't mention the relation between EEC and Norway for example, where Norway has a basic industrial fishery. EEC countries, except for Denmark, are basically interested in food fish. Since these are the same stocks, they interact and generate all sorts of problems.

There is also the question of distant water fleets, the tuna fisheries throughout the world oceans, and some other areas beyond EEC jurisdiction. Another common element is the question of effort control.

A third area of comparison is what one might call management goals in the US. William Gordon discussed optimum yield, as did Parzival Copes. It is clear in the EEC there is very heavy weighting for economic management.

That in fact brings up a fourth area, because that involves the trade office in fishery management. It is pretty clear from Copes' presentation that Canada is willing to trade off access to its fishery areas for a market interest and, of course, that is true in the EEC as well. In the United States we don't do that.

I suppose the fifth common point wasn't really addressed by anyone in any great detail. That is the question of basic fishery mangement. Jacob Dykstra mentioned it. He indicated that most bottom-fish fisheries and many surface fisheries as well are mixed species fisheries, and the theory and models we have for fishery management aren't very accommodating to that.

Comparative policy studies ought to think about across-the-board problems. A first one concerns basic data checks and the system that one generates for the collection, transmission, and utilization of data. And what about research concepts? This was mentioned briefly. There are many areas here. One concerns planning for fishery management; the other is the kind of research that deals with making beneficiary management plans in these areas. This basically lies in three areas. One is recruitment: what generates the variability in fish stocks. The second is multiple species interaction. The third is the interrelationship between the biological, economic, and social factors.

Another area concerns rules and criteria for management, like optimal yield and maximum sustainable yield.

The fourth problem is how to monitor the effects of fishing in ways that don't involve data. And what about enforcement regimes?

Many chapters here address the rebuilding of stocks. While this is a worthwhile activity, Mother Nature at times causes stocks to go into a down turn and many stocks don't have the capacity to "recover." We really don't know whether this is the effect of fishing or not. The Japanese Sardine used to have a fishery of about a million-and-a-half tons. The fishery disappeared. The catch was zero. No more Sardine today? They are now catching about two million tons of Japanese Sardine. In contrast take the California sardine, a very big fishery. It also collapsed. There is virtually little fishery today and so the fishery never recovered. We need to be careful about rebuilding stocks.

Another myth—perpetuated in the negotiating text of the Law of the Sea—is that allocations to departed fleets eliminate the waste tank in the stocks. It really depends upon what one is talking about.

We hear a lot about managing fisheries. We hear about multiple species fisheries, multiple time frames and ecosystems management. These are all hints in the right direction, but I think we have to realize that if we are going to use these fancier styles of management, we are going to have to acquire lots more data, lots more information and a lot more concepts than we already have.

Finally, I want to discuss the Third World. The extensions of jurisdiction around the world have created a new regime. The Food and Agriculture Organization and the community of fisheries assist developing countries with managing their stocks. This is a good thing for the United States, for Canada and the EEC, and for all the other countries involved, because really what most people are interested in is commerce for fish, and these activities are beneficial.

SIX

Marine Environmental Protection

Eric D. Schneider
University of Rhode Island
US Environmental Protection Agency

CHAPTER 14

Marine Environmental Protection in the Scandinavian Countries

KLAVS BENDER

Agency of Environmental Protection
Denmark

INTRODUCTION

The subjects I would like to consider will be the following:

1. A description of the institutional structures dealing with marine environmental protection in the Scandinavian countries.
2. The policy in the Scandinavian countries concerning ocean dumping of waste.
3. Policy and legislation concerning pollution from off-shore oil and gas production.
4. Pollution problems in relation to sand-and-gravel-exploitation from the seabottom.
5. Legislation and policy concerning pollution from ships.
6. Policy concerning municipal waste water treatment plant.

I will first try to define the Scandinanvian countries which are discussed in this chapter. We have five countries, which we call the five "Nordic" countries, namely Iceland, Norway, Sweden, Denmark and Finland. In the normal daily dialect in my country, we call Norway, Sweden and Denmark the "Scandinavian" countries. In the following it is those three countries I am going to consider. The greatest weight will be on the Danish system, which, of course, is the system I know best.

DENMARK

Denmark has a Ministry of the Environment, built up around the minister and five big agencies. The agencies are as follows: (1) The National

Agency for the Protection of Nature, Monuments and Sites; (2) The National Agency of Environmental Protection; (3) The National Agency for Physical Planning; (4) The National Forest Service; and (5) The National Food Institute. The total employed in the ministry is about two thousand. Three of those five agencies work with marine problems. The marine pollution problems are dealt with in the National Agency of Environmental Protection, but the Agency for Protection of Nature, Monuments and Sites deals, for example, with the legal and practical problems in relation to sand-and-gravel exploitation from the sea bottom. The National Food Institute deals with such problems as heavy metals and organohalogens in fish caught for consuming.

Inside the National Agency of Environmental Protection, the Marine Division deals with most of the marine problems, but, for example, municipal-waste-water treatment is dealt with in another division, and all grants in relation to industry are dealt with in another division again. It can cause some problems that several aspects of marine pollution problems are dealt with in different divisions, and in the end the director and the deputy director of the Agency of Environmental Protection are the persons who decide which policy the agency will follow. The director has reference to the Permanent Undersecretary of State who again has reference to the Minister for the Environment.

As far as I know the marine pollution problems in many countries are dealt with in connection with fishery problems, that is to say, under supervision of a fishery department. If I at this stage should summarize our experience and advantage of dealing with the problems under the administrative structure we have, I would claim that one can solve the problems in close connection with the solution of pollution problems from industry and pollution problems from municipal sources. One sees the marine pollution problems on a broader scale when one realizes that the input comes from sources like air pollution, land-based pollution, pollution from ships, pollution from ocean dumping, and so on. The system is organized, however, in a way to see marine pollution problems as one corner of the pollution problem of the whole ecosystem.

NORWAY

The systems in Norway are built much in the same way as in Denmark, with a Ministry of Environment built up of five big departments. Under the department for pollution is "Statens Forurensningstilsyn," which is comparable to the Danish Agency of Environmental Protection. The Norwegian Agency for Environmental Protection is built up of the following: (1) an administrative department; (2) an industrial department; (3) a department which only deals with oil pollution problems; (4) a department for toxic substances, ("produktkontrolafdelingen" is the

Norwegian name); and (5) a department for municipal waste treatment and recipient monitoring. The whole organization of dealing with marine pollution is very comparable to the system we have in Denmark.

SWEDEN

In Sweden there is no distinct ministry for the environment. The Ministry of Agriculture is the responsible authority on the governmental level, but in practice the National Swedish Environmental Protection Board is the central administrative authority when dealing with environmental problems. The National Swedish Environmental Protection Board also is built up of five departments: (1) Administrative department; (2) Department of natural resources; (3) Technical department; (4) Research department; and (5) Department of environmental hygiene.

The marine pollution problems are mainly dealt with in the technical department. The technical department is responsible for water protection, for matters concerning states' grants for municipal treatment plants, for environmental protection methods in industry, and for marine-pollution problems and oil-spill-combatting matters. In practice, Swedish Customs is responsible for oil-pollution control along the coast and on the sea; it is the organization which has the ships and the oil-combatting equipment. In this special aspect the Swedish system is closer to the US Coast Guard system than to the system we have in Denmark, where all kinds of oil-pollution problems and the combatting of oil spills are dealt with inside the marine division in the Agency of Environmental Protection.

SIMILARITIES AMONG SYSTEMS

If we should try to define some similarities in the Scandinavian systems, the most striking feature is that in all these countries we have marine-pollution problems dealt with in connection with land-based-pollution problems, air-pollution problems and other kinds of pollution problems. In the United Kingdom, for example, the marine pollution problems are divided between the Ministry of Environment and the Ministry of Agriculture, Fisheries and Food. As far as I can see, it is an advantage to look on the total pollution problem under the same authority.

All the Scandinavian countries have signed and ratified the international conventions concerning ocean dumping of waste: the Oslo and London Conventions. In addition to that, Sweden and Denmark are participants in the Baltic Convention, too, and those two countries have ratified the convention. But it is still not in force; West Germany and Poland have, for the time being, not ratified, and the convention will first be put into force when all the signature states have ratified. In the area which is covered by the Baltic Convention, every kind of ocean dumping of wastes is—except for dredge spoil operations from harbors—forbidden.

In the Oslo and London Dumping Conventions there are two annexes. In Annex I—which differs a little bit in the two conventions—are listed substances and materials which are totally forbidden for ocean-dumping disposal. Annex II is a list of substances and matters which require a prior, special permit for dumping in the sea. Of the three Scandinavian countries, Norway and Denmark have built their legislation in accordance with the philosophy of the conventions. But in both countries, dumping of wastes in the sea is looked upon as the last possibility for getting rid of wastes. We are, in other words, very restrictive with permissions. In practice, Denmark has only had one dumping of industrial wastes in recent years. Permissions have been renewed every year in the last seven years, and the issue is fully in accordance with Annex II of the conventions. In Norway and in Sweden they have—in the years after the conventions were ratified—never had any dumping of industrial waste. And in Sweden they have a law which goes further than the convention, totally forbidding any kind of dumping of industrial waste.

In all three countries one needs permission for a dredge-spoil operation. The administrative practice in my country is that—before we give or reject permission—we ask for the physical-chemical composition of the waste—and no more. In other words, we do not test the toxicity of the waste according to bio-assay tests. It is not our impression that bio-assay procedures are necessary according to the administrative procedure in connection with dredge-spoil operations. In these aspects we have—according to my opinion—less restrictive regulations than you have here in the US.

But it is difficult to be more precise here, because we have both in Denmark and in Sweden followed some single dredge-spoil operations more closely, with monitoring directly in the environment.

Incineration at sea is today looked upon as an aspect of dumping of waste in the ocean. Last year, the majority of countries in the London Dumping Convention accepted an amendment to cover this activity. Basically, the opinion in the Scandinavian countries was that this activity should have been totally forbidden on the sea. We preferred and still prefer a system taking care of chemical wastes by land-based incineration activities, by chlorolyses, by treatment of wastes in relation to cement production, or the like. It is problematic that one needs to reserve large areas on the ocean for incineration operations.

ADDITIONAL PROBLEMS

Further on, we see problems in relation to control functions. Today this control is good enough – in the cases authorities hear about. But when an activity takes place far out on the sea, as these activities do, the standard control problem really is something to be concerned about. Another big problem with incineration at sea is the release of large amounts of (mainly) hydrochloric gases. In the Scandinavian countries we have problems with acidification of the lakes caused by airborne pollution. That's the reason why we see the need for treatment of gases.

A third problem is the destruction efficiency: Is 99.95 percent always good enough? Could every kind of liquid organohalogenic waste be burned? For example, how big must the concentration of the dioxin TCDD be in wastes composition, and how big the total amount of dioxin contaminated wastes, to be burned on one position? If something goes wrong in the burning process, there is always a little delay in time before the wastes feed streams stop, and the plume will either impinge upon the ship or the oceans, with a high concentration of unburned wastes. These things happened, for example, during the burning of the Herbicide Orange which EPA granted in 1977.

Many of these problems are not related to at-sea incineration only. But honestly, are we forced to develop new techniques for the destruction of chemical wastes in the same way as we are if these activities take place at land-based plants? When this operation takes place on land, the operator is forced by authorities to treat the gases from the incineration operations, and the possibilities for controlling operations are greater. Alternative environmental groups could assure themselves of what is going on, and make their own measurement and control. And after all, in democratic countries one has to take aspects like that into account.

Only Norway – and to a lesser degree Denmark – have among the Scandinavian countries off-shore oil production. The Paris Convention – which is a land-based marine pollution convention – has recommended that its member countries accept a forty-ppm limit for concentration of hydrocarbons in the waste water, which is led from the off-shore structure to the sea. The waste water is mainly composed of production water, and in some cases displacement water. Nearly every oil field in the North Sea can meet these requirements; the Norwegian oil fields and the Danish oil field can. The total amount of hydrocarbons going to sea from our oil field – the Dan-felt – is between one-half to one ton of hydrocarbons per year from this sources. But this field is only a little oil field, with a total production of oil in the order of 3,500,000 barrels per year.

The problem for fishing activities inside the Oslo Convention Areas in relation to disposal of pipes, metal shavings, and other materials resulting from off-shore hydrocarbons exploitation and exploitation operations has been discussed for several years. The Convention recommends now that appropriate authorities of the contracting parties take all necessary steps to reduce the dumping of bulky waste from such activities into the sea. The policy in Norway and Denmark has been to avoid – as much as possible—problems of this kind in relation to off-shore oil-and-gas activity. Especially, Norway has very stringent regulations for the operators in force, regulations which, for example, require that bulky waste be stored and transported in containers for later land-based treatment. Further on, the regulations require that the licensee's name be on the container, and that large objects which are unsuitable for container storage be marked with the licensee's identity. They require that any loss of bulky wastes being transported shall be reported immediately to the national authority. And, after the licensee has finalized a drilling operation or a pipe-laying operation, there must be an inspection on the seabed around the site where the operations have taken place. If such an inspection shows any kind of bulky waste, the licensee may be required to remove the waste. We have only one operator in the Danish part of the North Sea. In practice, we have claimed the operator for mainly the same requirements.

Sand and gravel exploitation from the sea bottom is, in Denmark, regulated through the Raw Material Act of 1977. The aim of this act is to ensure that exploration of the Danish raw materials be based on overall planning and overall evaluation. The main problem for us is that our land-based resources in this respect are very limited. We have to change our production of sand and gravel materials from land to the sea bottom, and this shift will need this kind of regulation. The situation in 1976 was that production from sea materials counted for one-sixth of the total Danish production of stone, gravel and sand. The biggest problem in relation to an enlargement of the exploitation of sand-and-gravel resources on the sea bottom is with fishery. As far as we can see, the pollution problems involved in relation to this activity are limited, but we hope to gain a little more experience in this respect through a research project set up for the next two years.

SHIP-BASED POLLUTION AND WATER TREATMENT

Concerning ship-based pollution, the situation in the Scandinavian countries is as follows. None of the countries has ratified the MARPOL* 1973/78 convention. But, as far as I know, none of the countries with big ship tonnage have done that so far either. One of the problems in our country is the regulation concerning construction of new tankers; another problem is with reception facilities in ports. For the time being, only our big ports have reception facilities. The present situation is that legislation necessary for a Danish ratification of the MARPOL 1973/78 convention will soon be put into force. This legislation is also necessary

*Intergovernmental Maritime Consultative Organization MARPOL (Marine Pollution).

to fulfill requirements of the Baltic Convention.

Concerning municipal-waste water treatment, there are different philosophies in the Scandinavian countries, especially between Sweden and Denmark. In Sweden, discharge of sewage is regulated according to the environmental protection act which came into force in 1969. After this law, it became totally forbidden to discharge sewage to rivers or to the sea after primary treatment only. The normal situation in Sweden today is that all sewage discharged into lakes, rivers and the sea from urban communities should be treated both biologically and chemically after the primary treatment. Only in a very few cases has the authority accepted biological or chemical treatment alone after the primary treatment.

In Denmark we have quite a different philosophy. We issue grants to sewage treatment facilities according to the assimilation capacity of the receiving water. In other words, if the recipient in our country is a lake, the waste-water treatment plant must meet very stringent requirements. But if the recipient is the open sea, it is normal that we require only primary treatment. The exact requirement depends on the monitoring, which has to be done in the receiving waters.

FINAL CONSIDERATIONS

According to industrial treatment, our policy is that all treatment of the so-called Annex I substances—like mercury, cadmium, and organohalogenic substances—should be done with best technical means available, and be done in the factory before the waste water goes to the public system or to the sea. In practice, many Annex II heavy metals are treated in the same way, because it is impossible to separate them in a waste stream from an industry which has a high concentration of cadmium in the waste, for example. But in theory all Annex II substances are treated according to the assimilation capacity of the recipient. This philosophy is best seen when we examine discharges from the pulp paper industry, where the main problem is the very high content of organic substances. If monitoring activities in the waters where a discharge from a pulp and paper factory takes place are not able to show any disturbance in the natural bottom communities, we do not claim any special treatment of the wastes according to the biological-oxygen-demand content. The weak points in our philosophy are revealed through a question like: What *is* the natural situation, with animals and plants, for an undisturbed bottom community in a given area? A rather big and rather essential question, I must confess.

CHAPTER 15

Marine Environmental Protection in the United States

SARAH CHASIS

National Resources Defense Council
United States

The ocean serves as a sink for much of the United States' wastes. Rivers flow into the ocean carrying municipal, industrial and agricultural wastes. Pollutants are discharged directly into the ocean by shore-side communities, by vessels, and by offshore drilling. Elimination of these pollutants from the marine environment is one of the major goals of marine environmental protection. A related and equally important goal is the protection of the renewable resources of the ocean, the coastal and marine habitat, and the processes on which these resources depend.

When I speak of the marine environment, I refer to the coastal zone estuaries and the resources, waters, and submerged lands of the continental shelf and open oceans. Such a broad definition reflects the fact that these regions are environmentally related. Activities in one of these regions have profound effects on the other regions.

In the United States, the major sources of ocean pollution—exclusive of marine transportation and OCS* development—result from land—use practices. These activities yield pollution in the form of urban and agricultural runoff and direct discharges from the concentrated industrial, domestic and energy facilities along the coast. In addition, land-use practices affect the marine environment by modifying the habitat on which marine species depend.

Absent stringent control measures, pollution of the marine environment from land-based sources threatens to increase. Already over sixty percent of the United States' population lives in counties bordering the United States' shoreline. By 1990, this figure will climb to seventy-five percent. This population concentration will create additional sources of

*Outer Continental Shelf

ocean pollution, and pressures to modify coastal habitats. Coastal urban areas have already claimed extensive areas of productive coast. For example, forty percent of the original San Francisco Bay has been "reclaimed" by landfill. The Army Corps of Engineers' proposals to widen or deepen eleven major port channels will generate huge volumes of dredged spoil which—absent some alternative—will be dumped in the ocean.

Tremendous volumes of pollutants are already discharged into coastal waters. In New York alone, over two-hundred fifty million gallons of raw sewage are dumped daily into the Hudson and East Rivers. Nationally, one-sixth of the sludge generated by the treatment of sewage is dumped into the ocean.

Offshore as well as onshore activities will also be placing increased pressures on the marine environment. The United States government's program to lease over a million acres each year of the Outer Continental Shelf for oil and gas development increases the risks of damage to marine and coastal resources from catastrophic and chronic oil spills, and from the onshore impacts flowing from the siting of onshore support facilities. The increasing tanker traffic which is accompanying our increased consumption of imported oil also poses the constant threat of oil pollution, as demonstrated by the seventy million gallons of oil discharged off the coast of Brittany by the *Amoco Cadiz.*

The authority to regulate the resources of marine pollution and to protect marine resources is at present spread between all three levels of government in this country—federal, state and local. The states posses certain sovereign powers, including the power to regulate activities within their jurisdiction for purposes of public health, safety and the general welfare. The states' zoning powers founded upon these police powers are often delegated to local governments. As a result of this delegation, a majority of land-use decisions affecting the marine environment are in the hands of local governments. In contrast, most decisions involving events on the waterside are made by the states and federal government. The states exercise certain controls over the territorial seas and full ownership over the submerged lands out to the three mile limit. The federal government has exclusive or pre-emptive admiralty and interstate commerce jurisdiction. These powers limit a state's exercise of authority even within the territorial sea. Beyond the territorial sea, the federal government exercises control. For pollution prevention and resource protection purposes, this authority extends to the two-hundred-mile limit, or further, if drilling under the OCS Lands Act occurs beyond the two-hundred-mile limit.

Thus in looking at the protection of the marine environment in the United States, we must consider not only the extent and nature of federal authority, but also that of the coastal states, and in turn, the local governments to whom the states have delegated a great deal of land-use authority. Here I will focus primarily on the federal governments' efforts to protect marine resources, with some attention to the role of the states in this effort. I will first focus on those federal laws directed toward marine pollution control—What their major premises are, who implements them, and what some of the problems are. Then I will turn to those statutes directed more specifically at resource protection and conservation.

The Congress of the United States acted early, using the commerce clause of the Constitution to regulate and protect public waters. Congress passed the first pollution statute—the Act of June 29, 1888—in order to bar discharges into New York City's harbor. An expanded version of the 1888 Act—the Rivers and Harbors Act of 1899, otherwise known as the Refuse Act—outlawed dumping of refuse material into any navigable water or its tributary. This Act served as one of the main statutes protecting coastal waters until 1972.

FWPCA AND CLEAN WATER ACT

The most dramatic step forward in federal control of marine pollution came in 1972, with enactment of the Federal Water Pollution Control Act Amendments (FWPCA). The FWPCA regulates discharges of pollutants from point sources, such as power plants, municipal sewage-treatment plants and agricultural feedlots. It regulates oil and hazardous substance spills. It provides financial assistance for sewage treatment plant construction. Other provisions regulate vessel sewage, dredged spoil disposal, and non-point source pollution.

Discharges of pollutants from point sources are subject to both effluent limitations, which are limitations on the type and amount of pollutants based on availability of pollution-control technology and water-quality standards (minimum requirements necessary to sustain various uses of water). Water-quality standards are set by the states and approved by the US Environmental Protection Agency (EPA). Direct discharges conform to both effluent standards and water-quality standards, whichever is stricter. The basic enforcement mechanism is a permit system. Administrative and judicial remedies are available if a permit is violated. States with permit programs meeting federal requirements may assume authority to issue federal permits subject to EPA's veto.

Spills of oil and hazardous sustances are also regulated under the Act. EPA must designate a list of hazardous substances and define what constitutes a harmful discharge of both oil and hazardous substances. Any person in charge of a vessel or facility from which there is a harmful discharge of oil or a hazardous substance is required to notify EPA or the Coast Guard immediately. The federal government may then act to clean up the spill and mitigate its effects. The costs of these operations are chargeable to the discharger (with certain exceptions). Dischargers may be fined both for failure to give the required notice and for the discharge itself.

The FWPCA also establishes a separate permit system for the discharge of dredged or fill materials in waters of the United States, including coastal wetlands. These permits are issued by the Army Corps of Engineers, subject to EPA guidelines.

There is a specific section of the Act which relates to ocean discharges. This section (S403) requires the EPA to issue guidelines for assessing the effects of ocean disposal of pollutants and the availability of other methods of disposal, including land. Permits for ocean discharges must conform to these guidelines. There is an exception; ocean dumping from vessels in the territorial sea or further seaward is not subject to these guidelines. Such dumping is subject to a different set of statutory requirements, under Title I of the Marine Protectin, Research and Sancutaries Act of 1972, otherwise known as the Ocean Dumping Act.

In 1977, the FWPCA was amended by the Clean Water Act. The deadline for achieving the required level of pollution-control technology was extended, as was the deadline for the control of toxic substances into waterways (until 1984). Among other changes, the pollution-control zone was extended two hundred miles to insure the protection of resources—such as fisheries—under the exclusive management authority of the United States.

Full implementation of the Clean Water Act will go a long way toward eliminating marine pollution. However, implementation of this complex water-pollution-control scheme has been plagued with problems. Deadlines have not been met by the EPA. Numerous court challenges to EPA's effluent limitations have been brought. Hearings on permits drag on for years, and the challenged conditions remain ineffective in the meantime. EPA permits have become in some instances, permits to pollute. The control of non-point source pollution from urban and agricultural runoff has been an elusive goal. The Act thus has a long way to go before its goals are satisfied and the marine environment is adequately protected.

OCEAN DUMPING ACT

Another statute referred to before—the Ocean Dumping Act—authorizes the EPA to issue permits for ocean disposal of non-dredged waste material (such as sewage sludge and industrial wastes) and authorizes the Corps of Engineers to issue permits for ocean disposal of dredged material. The US Coast Guard has the responsibility to enforce the Act. Ocean dumping is prohibited unless authorized by permit, and certain types of material may not be dumped at all. (These include radiological, chemical and biological warfare agents, as well as high-level radioactive water.)

Permits for other substances are granted only if it can be determined that "such dumping will not unreasonably degrade or endanger human health, welfare or amenities, or the marine environment, ecological systems or economic potentialities."

At present dredged material is the most significant material being dumped in the ocean. (It accounts for ninety percent of the total waste disposed of in the marine environment.) A certain portion of this dredged spoil is contaminated with heavy metals and other toxic pollutants. Serious questions have been raised (by environmental groups) about both EPA's and the Corps' regulation of ocean dumping, particularly of dredged spoil material. These questions relate to whether:

> (1) ocean dumping criteria for dredged materials comply with the requirements of the Ocean Dumping Act or the international convention on ocean dumping;
> (2) whether the issuance by the Army Corps of Engineers of "interim" approval for one-hundred forty ocean dump sites without conducting any preliminary studies is a violation of the law or convention;
> (3) whether adequate consideration or planning has gone into alternative disposal networks, such as land disposal or the formation of contained offshore islands.

At present, under the Act and regulations, ocean dumping of all materials is to end by 1981. It appears, however, that this deadline will not be met. Ocean dumping of industrial and sewage wastes and dredged spoil thus continues to pose a threat to the marine environment.

OUTER CONTINENTAL SHELF LANDS ACT

In September of 1978, Congress passed amendments to the Outer Continental Shelf Lands Act, some decisions of which are geared specifically toward safeguarding the marine and coastal environment during offshore oil operations. There are provisions for the following: the suspension and cancellation of a lease, if operations could cause serious harm to the marine or coastal environment; the conduct of an environmental studies program; new safety procedures, including a requirement that the best available and safest technologies be used in offshore drilling operations; state and public review of plans for exploration, development, and production. These amendments represent a victory for environmentalists, but, in order to ensure that the new provisions result in effective controls over offshore and onshore Outer Continental Shelf activities, the Department of Interior's implementation of the amendments and of the oil-leasing program must be vigilantly monitored.

For example, the Department of Interior still proposes to lease areas in the fishery-rich Georges Bank, despite grave uncertainties concerning the effects of oil and gas development on the fishery. A recent decision by the US Court of Appeals for the First Circuit, while vacating an injunction issued against this sale, ordered the Secretary of Interior to protect the fishery resources of the area and, if he could not, directed him not to lease.

PORTS AND WATERWAYS SAFETY ACT

The Ports and Waterways Safety Act, which passed in 1972 and was substantially amended in 1978, provides important new protections for the marine environment. It requires that the Secretary of the agency in which the Coast Guard is operating set minimum standards for crude oil and product carriers. These standards include required use of segregated ballast tanks on new crude oil tankers of 20,000 dead weight tons or more, and new product carriers of 30,000 dead weight tons or more. These and other requirements tract the IMCO* Tanker Safety and Pollution Prevention standards agreed to in February, 1978. The Act authorizes the Secretary to issue different, more stringent standards for vessels in domestic trade. To date, over the strenuous objection of environmental groups, the Coast Guard has not done this.

ADDITIONAL LEGISLATION

There is additional legislation affecting the marine pollution, including the Deepwater Port Act of 1974. There are also a number of important bills pending in Congress affecting marine pollution. One of the most important of these relates to legislation establishing liability and compensation for cleanup costs and damages flowing from oil spills and spills of hazardous substances. The latest bill on this subject is the "Ultrafund" legislation, to establish a system of notification, emergency government response, enforcement, liability and compensation for oil and hazardous substance spills. In essence, the legislation creates a system of strict liability for damage and cleanup costs resulting from such spills, sets liability limits, and establishes a compensation fund for damages above owner/operator liability limits. The fund of $1.625 million is established from appropriations and fees levied on oil, petrochemical feedstocks, and certain other chemicals. The fund may be used to compensate for cleanup and mitigation costs, for real or personal property damages, or the loss of opportunity to harvest marine life.

*Intergovernmental Maritime Consultative Organization

PROTECTION OF MARINE RESOURCES AND HABITATS

In addition to legislation directed at pollution prevention and abatement, there are a number of federal laws specifically directed at protection and conservation of the renewable resources of the marine environment.

Title III of the Marine Protection, Research and Sanctuaries Act provides for the designation and protection of especially valuable areas of the marine environment. The program has been in serious trouble since its inception. Although the Act was passed in 1972, only two sanctuaries have been designated to date: the Monitor—off of Cape Hatteras—and Key Largo Coral Reef. Several other areas are now being considered for designation off of California and Texas. The main reasons why the program is so far behind are limited funding and opposition from the offshore oil industry. Many citizens' groups have nominated marine sanctuaries in areas where offshore lease sales are proposed, such as the Beaufort Sea in Alaska and the Santa Barbara Channel in California, but the industry seems determined to prevent any marine sanctuaries from being created around oil-and-gas-producing areas. Just recently, commercial fishermen and conservationists in New England nominated the Georges Bank as a marine sanctuary. This nomination has triggered strong reaction because of oil and gas development in the same area. The outcome of this debate will say much about the federal government's commitment to protection of the renewable resources of the OCS.

The Fishery Conservation and Management Act of 1976 is designed to insure not only the development of domestic fisheries, but their conservation and maintenance. Quotas must be set based on principles of optimum sustainable yield, taking into account ecological as well as economic and social factors.

The Marine Mammal Protection Act is designed to protect the dwindling numbers of marine mammals. The tuna/porpoise controversy arose out of this legislation.

The Endangered Species Act applies to all threatened or endangered flora or fauna. The Department of Commerce has responsibility for protecting most endangered or threatened marine species.

COASTAL ZONE MANAGEMENT ACT

In recognition of the importance of the coastal zone and in order to slow the accelerating development and destruction that was occurring there, Congress passed the Coastal Zone Management Act (CZMA) in 1972. The CZMA is designed to encourage state and local governments to plan for and manage coastal development so as to protect vital resources and channel growth into areas which can best handle it. Federal funding, plus the promise that federal actions will conform to state coastal zone management programs, are the incentives used to encourage coastal states to participate. The states, for their part, must develop coastal management schemes which meet federal standards. (So far, thirteen of the thirty-five states and territories which border the oceans or the Great Lakes have developed such programs. Twenty-two other states and territories are presently developing programs which may or may not be accepted.)

However, in evaluating state programs against federal guidelines, these are the problems we confront:

1. The absence of protections for key coastal resources, such as barrier islands, dunes, estuaries, and fisheries.

2. A lack of enforceable standards governing development in and adjacent to wetlands, dunes, estuaries and other valuable coastal resources.

3. Lack of specific policies and standards governing use and protection, so that it is impossible to tell what kind of protection in fact will be afforded when implementation begins.

4. The exemption from environmental controls of prevalent and potentially harmful uses of coastal areas, such as agricultural, silvacultural, residential housing, and highways.

5. The absence of mechanisms to assess and control the cumulative impacts of development on coastal resources.

6. A lack of affirmative environmental planning, that is, the failure to state clearly in advance that certain types of uses will be directed away from valuable coastal areas to other areas more suitable for development.

Substantial strengthening of this program is necessary if valuable coastal resources upon which marine species depend are to be adequately protected.

CHAPTER 16

Marine Environmental Protection: A Canadian Perspective

JAMES KINGHAM

Environmental Protection Service
Canada

INTRODUCTION

The individual problems which confront Canadian marine-policy makers are no different in Canada than in other countries. Taken together, however, the combination of cultural, institutional, geographic and economic factors which make up the Canadian mosaic present very complex problems to marine-policy makers.

The Canadian land mass extends across four-and-one-half time zones and runs from 41° North latitude to within about four hundred miles of the North Pole. Canadians are a "melting pot" mix of cultures although, unquestionably, the predominant ones are those of the English and French. In the Arctic areas, the native culture continues to predominate and, because of special dependence on the natural environment, this factor must be taken into account in defining and implementing Canadian marine policy in northern areas. To a large extent, jurisdiction over activities which could affect the marine environment is shared by federal, provincial and territorial governments.

Despite these problems, Canadian marine policy implementation has been fairly effective and the water quality in and around Canada is fairly good for an advanced industrialized nation.

THE CANADIAN MARINE ENVIRONMENT

For the purposes of this chapter, the term "marine environment" will be used to describe that part of the water environment in which one might expect significant maritime activity, whether in salt water or fresh. This somewhat unusual definition is needed in order to take account of the penetration of the St. Lawrence Seaway into the heartland of Canada and the United States. This penetration creates its own special problems for marine-policy makers, because it brings shipping into freshwater areas which are not only used for recreational and fishing purposes, but are also in some instances used for domestic and industrial water supplies.

The west coast of Canada affords some of the most spectacular shoreline viewing on the North American continent. Heavily wooded forests rise from the seashore along the Coastal Mountains, and one of the main industries, the logging industry, creates some pollution problems, as many errant logs from the marine transportation of forest materials litter the shoreline. A profitable herring fishery operates along the west coast, as well as an important salmon fishery. These industries are threatened by the risk of major tanker spills, a risk which has been increased over the last few years by virtue of the transportation of oil from Valdez, Alaska, to Cherry Point, Washington. This transportation route brings tankers through Canadian offshore areas and through confined waters such as the Straits of Juan de Fuca, near Vancouver. (Fig. 1).

The Arctic presents its own special problems. Until a few years ago, these problems were not ones which seriously threatened the Arctic environment. Basically, the native peoples of the north were living in dynamic haramony with the environment. The discovery of oil in Alaska changed all this. A proposal was made that Arctic oil might be shipped to the Atlantic seaboard of North America by means of ice-breaking supertankers plying the "Northwest Passage" through the Canadian Arctic. The threat of a major oil spill—and what would undoubtedly be a very slow recovery of the Arctic area from such spilled oil—coupled with the fact that in many areas native people in northern Canada continue to survive by virtue of a hunting and fishing way of life which depends on that environment, persuaded the Canadian Government to propose extraordinary measures for the protection of the Arctic environment. Subsequently, the possibility of finding oil in the Canadian Arctic has lead to current exploration programs, both in the Arctic islands and in the nearshore and offshore areas of the Beaufort Sea. These activities also present threats to the Arctic environment and need to be very carefully regulated.

The Atlantic offshore area has, for the last four centuries, been an important and valuable fishery area. Now this area faces the spectre of serious pollution because of the current traffic in oil and other hazardous materials to the East Coast of North America. Land-based activity of a potentially polluting nature, coupled with the current activity to find oil off the east coast of Canada, calls for a coordinated approach to the multiple threats in this area.

The Gulf of St. Lawrence is an important area in Canadian marine-

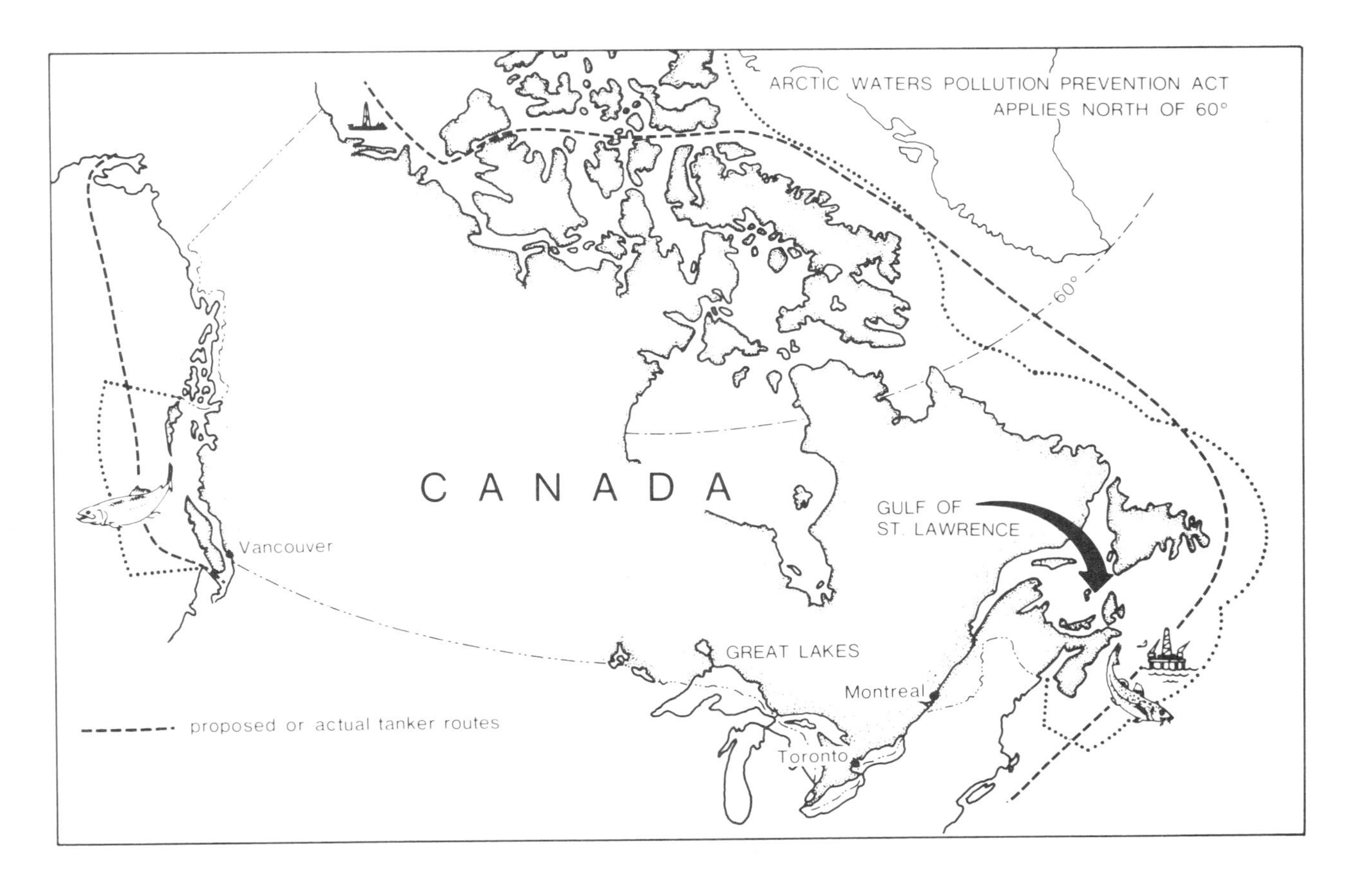

Figure 16.1: Canada—A Mosaic of Marine Policy Problems

policy considerations: it is the entranceway to many inland ports, and thus sees very diversified traffic; it is a major tourist recreation area with many fine beaches; and it is also a valuable fisheries area, both for fin fish and shellfish. The Gulf is ice-covered throughout the winter and this makes marine transportation hazardous, if not impossible.

From the Gulf of St. Lawrence, the St. Lawrence Seaway stretches some 1700 miles to Thunder Bay, Ontario. Along the route of this Seaway, ocean-going vessels traverse valuable near-shore shellfishery areas, pass by the two largest cities of Canada, travel over the largest freshwater reserves in the world (the Great Lakes system)—in which there is a thriving freshwater fishery—and pass over water intakes which supply drinking water to the most heavily populated area of Canada. Obviously, the standards for marine environmental protection, which would apply to vessels on the high seas, cannot apply in such confined and sensitive waters, and this calls for a special management regime, as well as bilateral (Canada/USA) cooperation.

BASIC CANADIAN MARINE POLICY

It is the policy of the Government of Canada to protect waters under Canadian jurisdiction from abuses which would: (1) pose a risk to human health or safety; (2) damage the flora and fauna living within, or depending upon such waters; and (3) reduce the value of such waters from an amenity point-of-view, or for other uses such as transportation and recreation.

This basic marine environmental protection policy is shared by all Federal departments and Provincial governments. As is obvious from the statement: it involves human health factors, and therefore the health agencies at both Provincial and Federal levels; it involves the protection of the fisheries, and therefore requires the involvement of Federal and Provincial fisheries officials; and it involves tourism, recreation, transportation, and other areas of interest to both Federal and Provincial levels of government.

This policy is implemented by identifying, for each of the sensitive areas under Canadian jurisdiction, the basic requirements for marine environmental protection. These requirements are then translated into Canadian law (in general, Federal law) which may be administered either by the Federal government directly or by delegation to the provinces (e.g., the administration of parts of the Canada Fisheries Act is delegated, in a number of instances, to provincial governments). Where a new threat to the environment is perceived, regulations or new legislation is drafted accordingly at either the Federal or Provincial level, depending on whether the threat is a general or regional one.

CANADIAN INSTITUTIONAL ARRANGEMENTS

Canada is a federation of ten provincial governments and two territorial governments under the wing of the Federal government. The powers of each level of government are set out in the British North America Act of 1867. Sections 91 and 92 of that Act assigned certain specific powers to the provincial level of government (e.g., education) and left the remaining powers (specified and unspecified) to the Federal government. The Federal government and each provincial government has its own Cabinet in which there are a number of ministers whose ministries are responsible for matters which may impact on marine policy and marine environmental protection. These ministries, at the Federal level, are primarily concerned with transportation, (since marine transportation is a Federal responsibility), environment, and fisheries and oceans. At the Provincial level, in some instances, there is a ministry responsible for fisheries, and there is in practically every province a ministry responsible for the environment. This arrangement is shown schematically in Figure 2.

Provisions are made either formally or on an ad hoc basis for meetings between the different levels of government. There are periodic first ministers' meetings where the Premiers of the ten provinces and the Prime Minister get together for general policy discussions on matters of national importance. At the Cabinet, or ministerial level, there are often meetings between ministers holding similar portfolios. An example of this is the regular meetings of the Canadian Council of Resource and Environment Ministers.

These meetings discuss specific problems, (e.g., shoreline preservation), and out of such meetings between federal and provincial ministers may come Federal-Provincial accords in a given subject area. The officials of the various line departments often collaborate on joint programs dedicated toward a common goal (e.g., the response to and clean-up of environmental emergencies, such as oil spills).

At the federal level, there are seven principal departments involved in marine-policy development and marine envrionmental protection. These departments are Treasury Board; Transport; Fisheries and Oceans; Environment; Indian Affairs and Northern Development; Energy, Mines and Resources; and External Affairs, as shown in the schematic, Figure 3.

As the name implies, the Treasury Board has a general responsibility for the disbursements of the Canadian Government. It controls the funds which will be spent for investigations of marine pollutants and the prevention of marine pollution. The Department of External Affairs has a general obligation to mesh national and international objectives in a diplomatic way.

The other departments have "line" responsibilities with regard to the marine environment which are directly related to their main objectives.

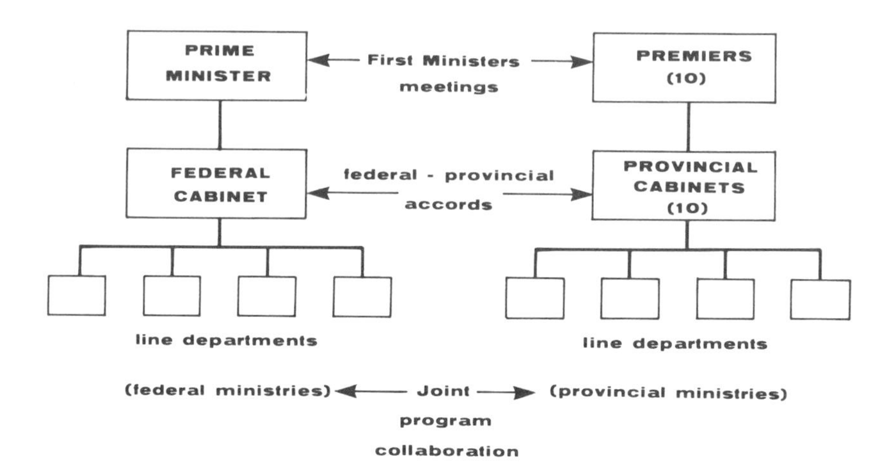

Figure 16.2: Federal-Provincial Cooperative Mechanisms

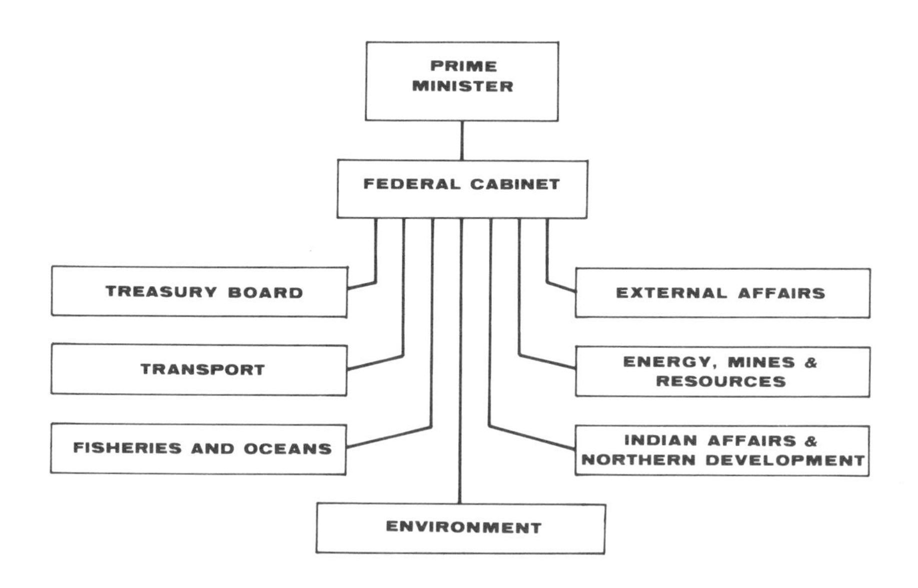

Figure 16.3: Federal Ministries Involved in Marine Policy

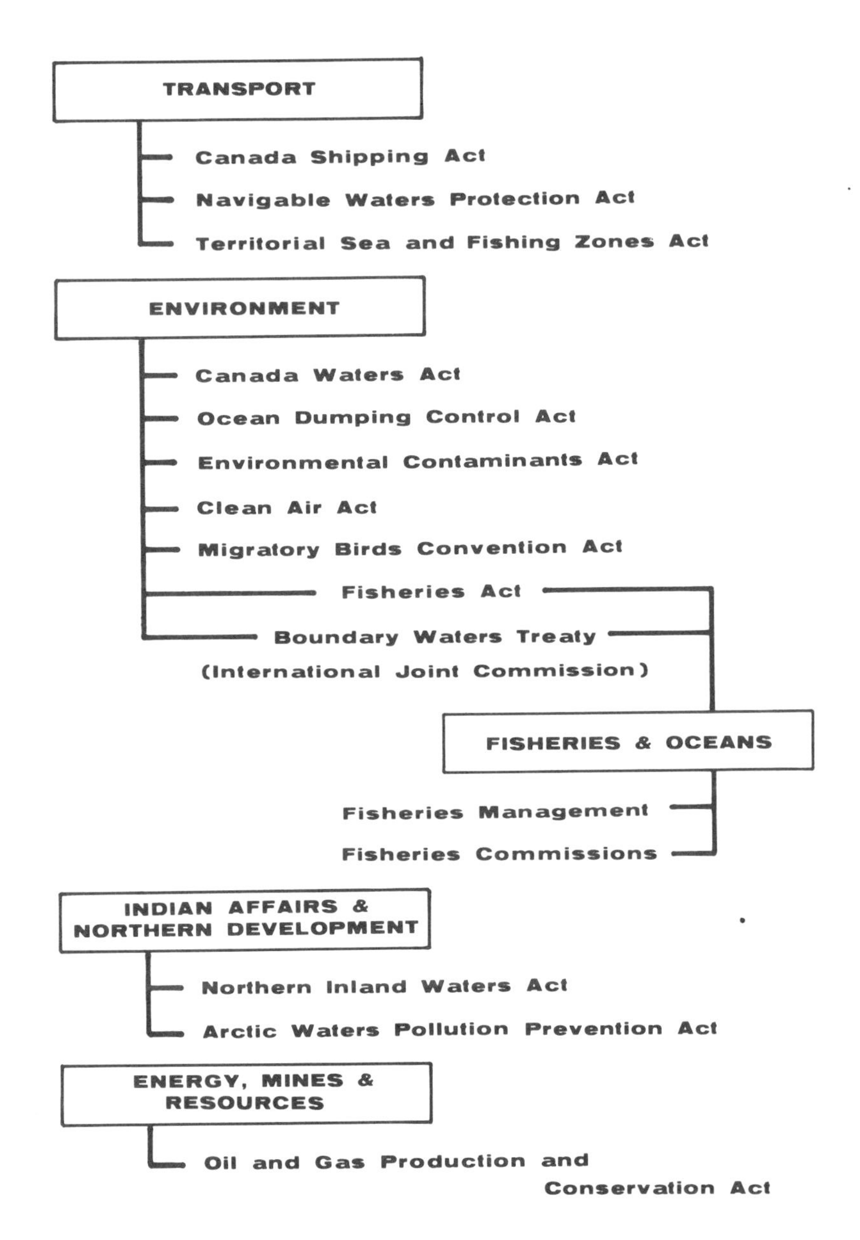

Figure 16.4: Some Canadian Federal Policy Instruments

For example, Transport is concerned with the prevention of pollution from ships and the maintenance of safe shipping routes. The Department of Energy, Mines and Resources is concerned with the prevention of pollution from offshore exploration or production activity. In the North, the Department of Indian Affairs and Northern Development is concerned with the protection of northern inland waters and the prevention of pollution of Arctic offshore waters, whether from land-based activity, from shipping activity, or from oil and gas exploration or production. The federal Department of Fisheries and Oceans is concerned with fisheries management, various fisheries commissions, and the protection of fish and the fish habitat. The Department of Environment is also concerned with the protection of the fish habitat because this constitutes part of the overall protection of the marine environment. In addition, the Environment Department controls ocean dumping, industrial contaminants, air quality, and boundary water quality.

CANADIAN LEGISLATION

Some of the Canadian federal policy instruments which are important in the field of marine environmental protection are shown in Figure 4. The following brief comments will illustrate how these various instruments are used.

A) *Transport*

1. The Canada Shipping Act, under Part XX, prescribes strict regulations concerning the discharge of operational wastes from ships. Provisions also exist under the Act for safe navigation of vessels.

2. The Navigable Waters Protection Act is basically an act designed to ensure that navigable waters are not interfered with by the unauthorized placement of barriers or other obstructions.

3. The Territorial Sea and Fishing Zones Act describes the limits of certain areas under Canadian jurisdiction and the activities which may or may not be carried out therein.

B) *Environment*

1. The Canada Waters Act establishes measures for the protection of Canadian freshwaters, not only from the point-of-view of fisheries protection, but also from the point-of-view of amenity considerations, recreation, and human health protection.

2. The Ocean Dumping Control Act prohibits the dumping of mate-

rial into the ocean except with the expressed permission of the Federal government, in accordance with provisions which are generally spelled out under the international London Dumping Convention.

3. The Environmental Contaminants Act prevents the introduction into the environment of materials which may be deleterious to human health and the environment. Although this is a general piece of legislation, one can see that certain of these contaminants could affect the marine environment if they were not controlled at source (e.g., PCB's).

4. The Clean Air Act is also a general piece of legislation which is intended to prevent the introduction of harmful substances into the air. It is a fact that the majority of pollutants (on the basis of total quantity deposited) which reach the oceans do so through the atmosphere. Therefore, it is important to prevent the introduction of dangerous substances into the atmosphere, as part of an overall marine environmental protection strategy.

5. The Migratory Birds Convention Act provides protection for birds and bird sanctuaries. In view of the fact that many of the birds thus protected are those which nest in marine areas, the protection of that habitat is a part of the marine environmental protection strategy in Canada.

6. The Fisheries Act sets out, among other things, measures for the protection of fish in waters under Canadian jurisdiction. While most of this act is administered by the Department of Fisheries and Oceans, parts of it, related to marine environmental protection, are administered by the Department of the Environment.

7. The Boundary Waters Treaty, and the International Joint Commission set up thereunder, is a mechanism for Canadians and Americans to identify mutual problems which may affect transboundary waters. Through this mechanism, these problems have been satisfactorily resolved from the point-of-view of both countries involved. A number of Departments, especially Environment and Fisheries, are involved in the administration of the provisions of this Treaty.

C) *Fisheries and Oceans*

1. The Department of Fisheries and Oceans is responsible for fisheries management activity. This includes the identification of quotas of fish to be taken in different areas, in accordance with the optimum sustainable yield in a given area. With the acceptance of fisheries zones which extend two hundred miles from national base lines, this fisheries management role has become increasingly important. Failure to properly manage fishing activities in the various areas under Canadian jurisdiction could result in serious ecological imbalances. Thus, fisheries management is considered here as part of an overall environmental protection program.

2. The Fisheries Commissions, which are also run by the Department of Fisheries and Oceans, are another important means of controlling fishery exploitation

D) *Indian Affairs & Northern Development*

1. The Northern Inland Waters Act is designed to protect freshwaters north of the 60th parallel of latitude in Canada.

2. The Arctic Waters Pollution Prevention Act is designed to protect the offshore Arctic waters from all sources of pollution.

E) *Energy Mines & Resources*

The Oil and Gas Production and Conservation Act is designed, among other things, to prevent pollution from offshore oil and gas exploration or production activity.

These Acts, taken together, constitute a considerable measure of marine environmental protection from practically every possible source. In fact, there are those who argue that in these difficult economic times the Canadian regulation of marine environmental protection poses an additional economic burden on Canadian industry. Rather than backing off on Canadian legislation, however, it would seem more appropriate from a global perspective to encourage all countries to have minimum environmental protection standards so that states which practice good environmental protection are not penalized in international trade vis a vis those states which degrade the global environment.

One way to reach the goal of global marine environmental protection is to adopt internationally accepted measures. Such measures have already been outlined at the Law-of-the-Sea Conference, in general, in articles 198 to 224. In particular, articles 208, 209, 211, 212 and 213 spell out the requirement for states, acting through competent international organizations, to make rules with regard to pollution from land, from the seabed, from dumping, from ships, and from or through the atmosphere. Should a Law-of-the-Sea convention finally be signed and ratified, then present Canadian legislative powers could readily be adapted to accommodate international conventions which may be developed under these specific 'protection of the marine environment' articles.

THE EFFECTIVENESS OF THE CANADIAN APPROACH

In general, the multiple approach taken in Canada has proven fairly effective in protecting the marine environment. There are no cases of outright degradation of that environment, and the fishery continues to be a viable industry. In general, the water is suitable for recreational use and clear for navigation and recreational purposes. The difficulties and possible duplication arising from more than one level of government being involved in marine environmental policy and marine environmental protection, constitutes a kind of "check-and-balance" in favor of the environment. Pollution havens are prevented by basic federal minimum standards and individual provinces may further enhance the quality of the environment by increasing the protection requirements in areas under their jurisdiction.

In summary, the multitude of Federal departments, and the two levels of government involved, are a necessary encumbrance in the Canadian system, but are not necessarily a negative factor from the point-of-view of marine environmental protection.

Commentaries

ROBERT LUTZ
Southwestern University
United States

INTERNATIONALIZATION OF NATIONAL APPROACHES

When in 1976 I wrote an article comparing national approaches to environmental problems, I noted the widespread acceptance that new institutions and laws were required to deal with perceived problems of ensuring environmental quality.[1] In another article about the same time, I suggested that such problems would increasingly be recognized as transnational in origin and effect, and therefore would call upon countries to devise approaches compatible with multilateral management of environmental concerns.[2]

Today, based on discussion by the many representatives of foreign countries presented here, and my own survey of recent developments,[3] it is possible to report that the direction of environmental management policies in other nations as well as the US has proceeded as predicted. Governments in their efforts to deal with myriad environmental problems have moved beyond the first stage of institution-building (recognition of problems and acknowledgment of need), through the second (trial and error of management approaches), and are now struggling with the third stage—implementation. Their efforts reflect a focus on the more difficult, longer-term problems. This has necessitated a reorientation of their institutions—a greater emphasis on prevention policies, the costs involved, and the integration of environmental considerations with other policymaking. In this stage of development, much greater attention is also being devoted to issues of transnational concern; to information gathering, development, and analysis; to technological development; and to the role of the citizen in the process of resolving transnational matters.

A somewhat unexpected development is the spate of regional and international responses to environmental problems. It is no fluke that many of these relate to marine concerns where, of course, transboundary environmental problems always abound. UNEP's efforts in the Mediterranean and with respect to shared resources, the North Sea Liability Convention, the IMCO efforts regarding tanker safety and oil spill liability, the Nordic and Baltic Sea Conventions, the Pollution from Land-Based Sources Convention, and even the UN Law of the Sea Conference, all reflect a trend to deal with marine pollution on a multilateral basis and suggest that national management of the marine environment is increasingly dependent on international approaches.

SPECIFIC ISSUE AREAS

Within the framework provided by these national and international developments are some key issue areas which deserve mention. As the procedural and substantive approaches for managing the marine environment mature, these areas of concern will occupy the attention of policy makers.

A. *Environmental Assessment*—The wide adoption of the environmental assessment technique demonstrates an emphasis on preventive approaches. Many countries, and regional and international institutions, have adopted the idea that a proposed activity can be assessed for its potential environmental impact. The types of activities covered by the requirement vary widely, but the interesting development here is whether the application of the advance assessment approach will be extended to cover extra-territorial projects having extra-territorial impacts.

B. *Information Development and Transfer*—For there to be any real achievement in the management of those environmental problems having transnational effect, it will be necessary for countries to develop techniques for sharing environmental information and for cooperating in its development. For example, crucial to dealing with long distant pollutants is the development of air pollution modeling techniques and the cooperation necessary for transnational monitoring. Overcoming national security concerns and guaranteeing the confidentiality of proprietary data are the principal problems here; they require very careful consideration for, at this time, they pose substantial impediments to meaningful exchange.

C. *Sophisticated Liability Questions*—Liability concerns seem to be moving away from the general question of whether there ought to be compensation for environmental harm. The new issues involve more sophisticated legal and policy questions. Three major foci can be identified. The first concerns the development of adequate standards of damage for environmental harm. The central issue here is whether standards ought to be developed which allow for the measurement of compensable harm to include ecological as well as the traditional economic value of resources. Methods for easing proof problems in establishing such value is the second focus. Standards which simplify the often imponderable causation requirements would recognize the complexity, interrelationship, and interdependency of factors contributing to environmental problems. This development, to some extent, relies on information development and transfer, referred to above. The third focus is on the problems associated with overcoming the barriers which often prevent adequate compensation for environmental harm. Legal approaches need to be developed and adopted to pierce corporate veils and to prevent instances where corporations are able to make themselves judgment-proof to avoid liability for environmental pollution injuries.

D. *Transnational Environmental Responsibility*—The frequently exhorted Stockholm Declaration Principles impose obligations on nations whose activities may have transnational environmental impact. Whether nations will respond to the extent suggested by those Principles and other customary international laws will depend on the flexibility and sensitivity of their environmental institutions to transnational problems. It is certain that the development of institutional sensitivity will require revision of jurisdictional, choice-of-law and enforcement approaches. Movement from adversarial and purely administrative (bureaucratic) decision-making to other forms of dispute-resolution and decision-making seem appropriate too. Standard-setting procedures must also be reassessed in order to be responsive to the concerns that standards are harmonious with the approaches employed by other countries.

E. *Citizen Participation*—It is axiomatic that the public should participate in the preparation and enforcement of major decisions regarding the environment. Although the citizens role in providing environmental quality is expanding in many countries, the extent of participation and the points in the decision process it is allowed continue as central issues to this development. Since the quality of input is often seriously hampered by the complexity of issues and the voluntary nature of many citizen efforts, questions of institutional support for citizen participation will be increasingly raised.

POLICY BARRIERS TO THE INTERNATIONALIZATION PROCESS

The satisfactory treatment of three policy concerns is instrumental to the development of effective national and international approaches for managing transnational environmental problems. The significance of these three is such that the failure of a nation or transnational institution to respond to any one of them may seriously impede its management efforts. The three potential barriers are: harmonization; economic cost-benefit of management techniques; and the role of developing countries.

A. *Harmonization*—Defined as the national adoption of laws similar to other countries and often referred to as "approximation of laws," this process is necessary in the transfrontier pollution context. The harmonization of standards between countries makes possible the transnational management of pollution problems and is also considered necessary to avoid what would otherwise be the economic "distortion of competition." The concept mandates the complicated and virtually impossible political task of multinational adoption of similar pollution standards. On the other hand, harmonization might be employed to only those problems which have a transnational and localized nature. A limited, but certainly viable approach, would seem to be the adoption of common standards and procedures specifically designed for and limited to frontier areas involved in a particular problem.

B. *Economic Cost-Benefit of Management Techniques*-Due to the world-wide economic situation, few nations can escape the dual problems of rising inflation and energy shortages. For environmental regulation, this has resulted in an active debate about relative costs and benefits. Many studies have been conducted, none of them conclusive, which try to associate the economic benefit from environmental regulation with the economic costs involved, or vice versa. These sorts of analyses divert our attention from the real challenge of insuring that economic and environmental problems are dealt with in harmony. Greater focus should therefore be placed upon the development of prevention techniques and on understanding the interrelationship of policies, including energy, and regional and industrial development.

C. *The Role of Developing Countries*—The role developing countries have to play in the management process has been omitted in our discussion so far, but their function is very much related to achieving international environmental quality. As already demonstrated by the efforts to obtain international consensus in the UN Law of the Sea Conference, the "New International Economic Order" (NIEO) poses an agenda of concerns affecting developed countries in their resource development and international environmental quality objectives. Despite the lack of internal discipline of the NIEO and the often untested allegiance of its members to the precepts of the NIEO, it has proven a major barrier to achieving the consensus necessary amongst nations in dealing with international environmental problems. Clarifying the relationship between developed and developing countries in any international resource management efforts depends on the resolution of two policy considerations. First is whether the "double standard," often invoked by developing countries, is a legitimate and workable requirement to environmental management. There is no doubt that achieving environmental quality imposes substantial costs upon a country. With developing countries, these costs can far exceed their ability to pay and can have a stifling effect on their economic growth. On the other hand, the lack of management efforts by some countries will certainly not help achieve international environmental quality. Closely linked to this first issue is the second, which focuses on the responsibility of developed countries towards developing countries in working toward environmental quality. If it is deemed appropriate, as the NIEO principles suggest, for developed countries to subsidize and lend assistance to developing countries' efforts, then the question becomes how much and in what circumstances.

CONCLUSION

I hope these remarks have synthesized some of the developments in the environmental management area and have served to identify some of the major obstacles to achieving transnational and international environmental quality. There is much ado about environmental management in the countries focused on by this conference. Many challenges remain, and for that reason I think it appropriate to conclude my remarks with a quote from UN Secretary General Kurt Waldheim, made on the occasion of World Environment Day, June 5, 1979. He called for a renewed pledge by governments and people everywhere to safeguard the human environment, and to improve the quality of life "in a world of equity at peace with itself and with nature."[4]

NOTES

1 See generally, R. Lutz, "The Laws of Environmental Management: A Comparative Study", 24 *American Journal of Comparative Law* 447 (1976).

2 See generally, R. Lutz, "An Essay on Harmonizing National Environmental Laws and Policies", Pt. 1 in 1 *Environmental Law and Policy,* 132 (1975), and Pt. 2 in 2 *Environmental Law and Policy,* 162 (1976).

3 See R. Lutz, "Directions of Environmental Law in the International System: An Assessment of Tasks and Challenges for Lawyers", in ENVIRONMENTAL POLLUTION AND INDIVIDUAL RIGHTS: AN INTERNATIONAL SYMPOSIUM (ed. S. McCaffrey and R. Lutz, (1978).

4 *New York Times,* (June 6, 1979), p. A5.

STJEPAN KECKES

United Nations Environment Programme
Geneva

I would like first to emphasize several facts that are not so clear in our discussions. The first is that the boundaries of marine pollution frequently do not coincide with national boundaries. You might not feel it so much in the United States because you are a big country, a continent, a world in itself. Smaller countries are more inclined to see marine pollution problems not as internal problems only, but as a matter requiring foreign policy considerations and decisions. Another fact, frequently forgotten, is the very distinct feature of developing countries in that their priorities look and are different from those of the developed countries. Their first priority is, naturally, development, and if you can try to induce them to do something about the protection of their environment, you have to present your case and you have to work in the context of their development strategies because this is their absolute priority. One very distinguished leader of a developing nation has said: "If environmental protection means a loaf of bread less for my people, I am against it". And they are definitely against it if they feel that environmental protection means the slowing down of their development. That is definite.

So we tried to approach the environmental problems of developing countries on their own terms. These terms can be defined very easily if you present environmental protection as the protection of man, the protection of the health of man and the protection of natural resources necessary for his economic development. Under these terms the developing countries are interested in environmental protection.

In a specific example I shall show you how it is done and how it influences the national legislation of various countries. I shall remind you of the great concern about mercury pollution caused by the massive poisoning in Minamata. When it became evident that the poisoning was caused by seafood, the World Health Organization recommended a limit around 0.5 parts per million of mercury and many countries in the Mediterranean—because the whole story is about the Mediterranean—followed this recommendation and adjusted their national legislation so that the national law permits seafood only if it doesn't exceed a mercury limit which was rarely set above one part per million.

At the time when these national legislations were made, in the early 1960s, practically no data on the concentration of mercury in Mediterranean seafood existed. But towards the end of the 1960s, we started to realize that the actual mercury levels in Mediterranean fish—particularly tuna and swordfish—are frequently much higher than national legislations would allow. Scientists first hesitated to publish their findings because they were afraid that the figures might not be correct, but also because they were afraid that it might cause trouble for them. Therefore, when in 1975 UNEP came on the scene, a massive research and monitoring program was initiated to clarify the situation. The results of this effort, in which thirty-six laboratories from fifteen Mediterranean countries participated, were surprising. The average mercury level in the Mediterranean tuna and swordfish was found to be between two and two-and-one-half parts per million, which is roughly three to five times more than the national legislation in most of the countries would allow.

When compared with mercury levels in Atlantic tuna these figures are very high. The biologists told us that the tuna in the Atlantic and the tuna in the Mediterranean belong to two different stocks which do not mix. We felt that the high Mediterranean figures cannot be explained by the pollution of the Mediterranean by mercury. Therefore we launched a thorough study of the input of mercury into the Mediterranean. The best estimates showed that the total input of mercury in the Mediterranean from land-based sources is about 130 tons per year. Of this amount about one hundred comes from man's activities, 90 per cent of it through rivers. This is natural, because the Mediterranean has quite a high content of mercury in the rocks surrounding its shores. Some of the oldest mercury mines are located in the Mediterranean region. In addition, you have a very active volcanic process, underwater and above the sea, going on in the Mediterranean, which is also adding a lot of mercury to the water. So it was understood that the high levels of mercury observed might have natural causes. We lacked more historic data. Tuna samples from museum collections have been analysed, although you have to be very, very careful interpreting these results because of the possibility that samples have been contaminated, or that they have lost part of their mercury content in the process of conservation.

We also looked at epidemiological data, concentrating on people dealing with tuna, fishermen in particular. We couldn't find any traces of sickness, illness or anything like the "mad Hatter syndrome," although some wags tried to say that *all* Mediterranean people are crazy from the mercury they are eating with their fish.

So we were faced with the situation that Mediterranean tuna and swordfish had up to five times more mercury than national laws would permit, and yet this is most likely a natural phenomenon which has existed as far back as we can trace it in history. The only conclusion we could come to was that there was nothing wrong with tuna, there was no serious pollution of the Mediterranean by mercury and there was nothing wrong with consuming such tuna.

We are now looking for various mechanisms to explain why the relatively high levels of mercury do not cause any measurable effects, and the specific metabolic pattern of selenium might provide an answer.

The facts were there and we had to confront the governments with them. The governments could only react in two ways: either ban the tuna from the Mediterranean kitchen or change their national legislation, as there was no evidence that consumption of the tuna led or leads to any harmful effects. They resorted to the second option. They started to change their national legislation so that it came into line with the facts. But our advice was: don't just relax the standard but educate those who will see to its enforcement. The standard was based on the assumption of a relatively high protein diet being supplied by tuna, which is not the case in the Mediterranean. So don't just relax the standard but link this standard to the consumption of seafood. Educate the population, put up signs on the fish markets saying "don't eat more than two kilos of tuna per week." This is the right approach to the problem. It is easy to legislate or to ban, but it is not so easy to instruct people how to live with certain things that might be hazardous. It might be a different approach from the one you have in the United States but I think that in regions like the Mediterranean we will have to resort to this sort of action more and more. Other alternatives do not seem reasonable. We wouldn't be able to justify them on scientific grounds, even less on economic grounds.

I have shown you in this case how we try to protect people without prohibiting them from doing things they should be allowed to do.

JAN PRAGER

Environmental Protection Agency
United States

I think it's not an accident that in discussing United States marine policy as compared to policies in the Scandinavian countries and in the Mediterranean, a number of people have commented that when it comes to environmental protection we seem to be a peculiarly litigious society. We appear to set marine protection policy in the courts. That may be because the nature of our marine policy in the United States is best described by its absence. We simply don't have a marine policy as such. Probably the closest thing to a policy is Section 101a of the 1972 Clean Water Act, which states, "It is our national goal to restore and maintain the physical, chemical and biological integrity of our nation's waters."

Every US regulatory official has read section 101a, but I have yet to hear the first lawyer use it in case arguments, either in an adjudicatory hearing or in our federal courts. I find that very interesting because it is a very clean-cut statement of goal. Unfortunately, it is thought of as an oversimplified goodness-and-motherhood statement at the beginning of an otherwise Byzantine law. That is too bad.

There is another reason why we are a litigious society. Our government has not been very successful in bringing about public participation in the development of its laws or in the drafting and approval of regulations. We have something called the "Administrative Procedures Act" which gives our public more than one opportunity to give advice to its government. But it has been used largely as a delaying tactic by people who deplore one government activity or another. Delay is used not only by environmentalists, as our industry representatives have pointed out. It also has been used by industry when a decision has not gone their way. The relationship between the Administrative Procedures Act (APA) and the National Environmental Policy Act (NEPA) has been of a delaying nature. I will be curious to see how our new regulations put out by the Council on Environmental Quality will affect the implementation of NEPA in the future. It is unfortunate that NEPA and the APA have been used only as a delaying tactic. They should be employed to improve the quality of federal decisions.

There is also the matter of public trust. Because of the recent history of our country—back to the end of World War II—the Executive Branch and the Legislative Branch of the US Government have not held the firm public trust they maintained during the preceding war years. In the post-war years, the Judicial Branch of the government has by and large enhanced its image and has merited more public trust than either of the other two federal branches. This too contributes to the litigious tendencies that seem so very peculiar to the US.

Cultural phenomena bear consideration in developing the future of our environmental movement legislatively, economically, politically, socially and scientifically.

Another comment I would like to make deals with linkages—linkages among scientific disciplines that are used in risk analysis—a very basic part of environmental management. People at the political level who make decisions really have no right to ask scientists anything more than, "What am I risking?," "How great is the risk?," or "What is the probability of occurrence?"

The problem with our present risk analyses has been that they are much too limited. Risk analysis requires a great deal more scope and range than we have employed thus far. Certainly in the Executive Branch of the Federal Government we need to look at whole systems. When we regulate ocean dumping of sewage sludge we should be considering, for example, the needs for fertilizer of our western timber producers, and farmers and the needs for economic stimulation of our railroads—not simply the harmful components of sludge and how many fishes it might affect adversely.

To some degree, risk analysis has suffered because our science still is primitive. There are some rather basic scientific problems which we really can't advise on very well. Concerning ocean dumping, the question is containment versus dispersal; which is better and under what circumstances? This very fundamental principle is still a matter of considerable argument. The issue of isolation versus destruction of pollutants is a related question that we have not settled as a matter of policy.

Concerning oil rigs offshore, there is a certain argument that claims they are sanctuaries for fishes, rather than hazards to them. I think that the validity of this argument needs to be examined carefully. Certainly when you read the advertisements, in the big color ad for the oil companies they will tell you that around every offshore drilling rig there are lots of fishes and that oil rigs provide the best fishing in the area. It is true. One needs to weigh the artificial reef aspect of oil rigs against the fisheries' production of the whole area. Again, I think that we need to look at whole systems.

Another major problem we deal with, at least within the US and on a national level, is fragmentation of decisions. A number of federal agencies contribute bits and pieces of major policy decisions and they are not very well-coordinated. Certainly the Canadian system as explained to us seems to foster better coordination than we have. Perhaps it doesn't. It is the nature of those of us who work for governments to try to defend them. It is my private nature to be an iconoclast and not to do so. In fact, the United States has a lot of trouble coordinating policy among agencies and among our three levels of government.

In conclusion, I submit that there is some hope for federal interagency coordination of ocean pollution policy in the United States. There has been a law on the books for almost two years now called the "Ocean Pollution Research Development and Monitoring Planning Act," which says that there shall be coordination of ocean pollution research in this country and that it shall be led by the National Oceanic and Atmospheric Administration (NOAA). Of course, the law goes on to say "this law doesn't affect those direct mandates given to other agencies or any ongoing programs." It also says that "NOAA shall come up with a five-year plan every two years forever." Nevertheless it is very promising. I am quite pleased with it. I think that it will bring our largest ocean agency with its great and excellent resources, people, and facilities into the ocean-pollution picture in a coordinating role. The first attempt at a federal plan probably will not be everything that all of us concerned would like it to be. But in two years we shall create another plan and two years after that still another one. Sooner or later we will do it right.

KENNETH KAMLET

National Wildlife Federation
United States

I would like to address briefly an issue that concerns general marine pollution control strategies and philosophies. Sarah Chasis has said you could expect me to say more about dredge spoil problems, and I will, but I would like to discuss those things in a wider context. As many of you, I am sure, know, the use of water quality standards has played a fairly prominent part in pollution control strategies in the United States and elsewhere in the world. More recently, technology—based pollution-control requirements have come to the fore and some combination of the two still continues to govern and dominate pollution-control strategies throughout the world.

One thing that has been really ignored, neglected very seriously in the course of this, is what could be referred to as sediment-quality standards—or an approach that considers the very significant role of bottom sediments, both in the ocean and in inland waters, not only as sinks for contaminants of various kinds but as significant sources of contamination of marine and aquatic food chains, including human food chains. That is where dredge spoils come into the picture, but the problem is not limited to dredge spoils.

Liquid wastes of various kinds which are discharged into waterways of various kinds—ocean and inland—often become absorbed into or otherwise associated with bottom sediment. The Kepone situation in the James River in Virginia, and in the Chesapeake Bay in Virginia and Maryland, is one example of that sort of situation. It is by no means unique. There, as most of you are probably aware, although one would be hard pressed—even while the Kepone discharges were still going on (but certainly after cessation of those discharges) —to demonstrate any increase in the water column, and yet fish and shellfish in those waterways were being contaminated with Kepone at levels dangerous to human health.

The consumption and the commercial taking of many species of fish and shellfish continues to be prohibited today because Food and Drug Administration limits for human health are being exceeded, even though the discharges of liquid Kepone have been stopped for several years now. Why is that? The reason is that the Kepone became absorbed into bottom sediment in the river and bay and continued—long after the liquid discharges ended—to serve as reserviors of contamination to overlying food chains.

Similar things have occurred in the southern part of California. DDT and PCBs were for many years discharged by point sources into the Southern California Bight. Those practices have been abated and very little in the way of liquid inputs of those material continues to go on. Yet, researchers out there continue to demonstrate significant contamination of bottom fish and shellfish. The only possible source of that contamination, again—bottom sediments.

The same thing has been demonstrated in the Hudson River and to a little lesser extent in the New York Bight. The Great Lakes have been having similar problems with PCBs—again deriving from bottom sediments in the absence of any known liquid point-source discharges of these materials. I find the implications of that rather disturbing because there is nothing in existing pollution-control strategies in this country—and in most ways this country is well ahead of other countries. If we have not addressed this, it stands to reason that most other countries in the world have yet to come close to addressing it. Nowhere in the world is there a systematic effort to take account of this problem of bottom-sediment contamination.

If we applied our traditional technology-based treatment controls or water-quality standards, particularly with the allowance they make for mixing zones and so forth, Life Sciences Company—the facility responsible for the Kepone discharges—would probably still be in business today. If this problem of contamination of aquatic life had not been detected, those discharges would still be going on. They would have the blessings of federal and state permits and no one would be aware there was any problem.

The testing procedures employed are not designed, plainly and simply, to pick up problems of that kind. And that concerns me a great deal. There are approaches that could be followed and I have proposed such approaches to the Environmental Protection Agency and to other agencies of the federal government. So far they have not been terribly well-received.

One approach would be to make use of a variant of solid-phase bioassay tests or benthic bioassay tests, which right now are being used solely to screen dredged materials proposed for ocean dumping in this country. But I would propose to see such tests applied to liquid wastes. How do you do solid-phase tests on liquid waste?

One way of doing that would be to do the toxicity test in an aquarium that contains not only the representative marine organisms whose sensitivity to the material and propensity to accumulate the material you want to look at, but also in the presence of bottom sediments of the sort that one would expect to find in the receiving waterway that is going to get the discharge. That would give one the opportunity to measure what is going on in a situation that resembles the real world. In some cases, for certain types of contaminants—perhaps for certain heavy metals—the presence of bottom sediment may reduce the toxicity to the aquatic organisms the toxicity test is performed on.

But for many other types of materials—chlorinated hydrocarbons seem to be the major group of chemicals in this category—it seems just as clear that the toxicity would be increased as a result of testing of this kind. This risk is especially acute where dredge material itself is concerned, because there you are taking bottom sediment and directly disposing of it.

In the case of ocean dumping in the marine environment or in other aquatic environments, you maximize the opportunity for these sediments to release their contaminants to marine food chains and other aquatic food chains. Yet, despite that potential, despite the evidence that I have briefly outlined for this as a serious problem, this country—the Corps of Engineers and the Environmental Protection Agency—and particularly other countries in the world, have tended to minimize the significance of the problem and have shown a great reluctance to do anything to constrain dredging operations or disposal of dredged material.

Klavs Bender indicated that the Baltic Convention prohibits all ocean dumping except for dredged spoil operations from harbors. He indicated that within his own country, Denmark, they require physical-chemical characterization of dredge spoils, "but no more," in his words. No bioassays, for example, are at all required in Denmark, and doubtless in most of the other countries throughout the world.

Within the United States, as Sarah Chasis mentioned, dredge spoils represent 90 percent of all ocean dumping going on off the coast of the United States. Not all of that material is contaminated with hydrocarbons and other toxics of concern. In many cases we don't have the foggiest idea what those bottom sediments contain.

Some preliminary analyses done by one of the National Oceanic and Atmospheric Association's laboratories from Seattle in the New York Bight area, of bottom sediment within the bight or tributary waters, indicate there are numerous exotic and highly toxic synthetic organic materials—chlorinated and otherwise—present in significant amounts. Nobody is significantly testing for those materials. Nobody is significantly evaluating the toxicity of those materials or their potential for contaminating marine food chains, and I think that is terribly unfortunate.

Let me get off my soap box on that score and pass quickly to another subject that Klavs Bender alluded to—perhaps to illustrate that neither the National Wildlife Federation nor myself are absolutely opposed to all forms of ocean dumping or ocean disposal.

I like to think that our position is one of rational waste management; that the ocean ought to be used as a disposal option of last resort rather than first resort, as it often is. There is always another lake or river to pollute. There really is only one ocean, and I think we have to think of things a bit more in that context. But Klavs Bender refers to the Scandinavian position on at-sea incineration. I recognize the concerns that he has raised and nobody is more concerned about the prospects of an incinerator ship loaded with Agent Orange herbicide, or some other similar toxic material, getting involved in a collision at sea and emptying its cargo into the ocean. Such a spill could devastate phytoplankton and other marine life for hundreds of miles, and that is a very sobering prospect.

On the other hand, one does need to look at the alternatives. Chlorinated organics, because they are so persistent, really can only be managed in environmentally-sound ways by destroying them by some thermal process. It is not the sort of thing that you want to put in a landfill, where there is a potential for it getting into surface waters or ground waters. It is not something you want to mess around with trying to recycle, because there are already too many of those things actively circulating in the environment today. So thermal destruction of one sort or another is probably the best fate for that material. The question becomes "do you want to incinerate that material on land or at sea?" Now Klavs has given very good arguments for trying to do it on land. I think the best argument for that is the facilitated opportunity for monitoring and keeping track of what is happening. An incinerator ship far out at sea, far from the prying eyes of a monitoring entity, may have the temptation to dump the material overboard rather than incinerating it. There are ways around that, I think.

There are automatic-sealed recording devices—black boxes, so-called—that can measure any unauthorized releases of material and can be inspected by regulatory authorities when the ship returns to port. I think there are ways of controlling that problem. There are a couple of very significant advantages, though, that at-sea incineration could have over incineration onshore for waste of that type. When you are out at sea and incinerating a chlorinated organic waste, the principal waste product that is generated in any sizeable amount is hydrogen chloride gas which, when it comes in contact with water, becomes hydrochloric acid—not the sort of thing you want to release into the atmosphere of a big city. If you did this onshore you would need to do it with very efficient stack-scrubbing devices to preclude the escape of HCl vapors into the surrounding atmosphere. That imposes constraints on the design of incinerators that have to have these stack control devices. It limits the size of the stack. It limits in other ways the design of the incinerator and in some ways thus reduces the potential to design the incinerator to produce the optimal degree of destruction of the material that you are trying to incinerate. The most complete destruction as possible is the objective that you are really after, and if an at-sea incinerator is capable of achieving greater destruction than an onshore one this is a very strong argument in my view for preferring the offshore approach.

There are other considerations involved here too. One, also relating to stack scrubbing devices, is that since HCl gas is highly corrosive, it would be very difficult to maintain stack scrubbers operating efficiently in that kind of hot, corrosive atmosphere.

The down-time would be very significant, and much of the time unscrubbed emissions would probably wind up being released into the atmosphere because the scrubbers were out of order or were not operating efficiently.

An incinerator vessel operating two-hundred miles off the coast would not be subject to that sort of problem, because there any released HCl can be assimilated by the atmosphere and marine waters. At-sea incineration takes the opportunity to make use of dispersion and dilution in one of the few instances where that kind of philosophy makes any sense.

SEVEN

New Directions In Comparative Marine Policy

Timothy M. Hennessey
University of Rhode Island

CHAPTER 17

Evaluating Marine Policy: Criteria From Two Models and a Comparative Study

TIMOTHY M. HENNESSEY

University of Rhode Island
United States

"First we had the Russians and now we have you. I prefer the Russians; there were fewer meetings."

Galilee, Rhode Island fisherman to a representative of the National Marine Fisheries Service.

"One cannot demand that decision-makers have perfect information as did the classical theory of the firm, for actual decision-makers never do. One cannot demand that they make impossibly elaborate calculations in making choices. The requirements which the assumptions impose must be to human dimension."

John Steinbruner, *The Cybernetic Theory of Decision.*

INTRODUCTION

In recent years comparative public policy analysis has received more attention in political science. Nevertheless, empirically based, comparative analysis of marine policy has been less frequent than one might otherwise have expected. I think Robert Freidheim's excellent review of the state of the art supports this generalization. These remarks are simply preliminary to the working assumption that much needs to be done in the analysis of comparative marine policy, and it is my hope that this forum and other forums like it will move the field forward.

In this chapter we argue—contrary to the received wisdom—that in the United States we do have a marine policy, or more properly policies. In fact, we do not have too little but too much. This is the case in the sense of our ability to implement the legislative mandates, given much of the current institutional structure, past "routines" of behaviors, and, above all, given our quite inflated view of what large-scale complex bureaucracies can do. Indeed it is precisely our understanding of the design of institutions and the relationship to decision-makers which is the central issue. I shall attempt to address this in my chapter.

Only ten years ago the Stratton Commission issued a blueprint for United States oceans policy. Since that time the National Oceanic and Atmospheric Administration, the Environmental Protection Agency and other federal agencies have been charged with carrying out the following oceans-related legislation:

> The National Environmental Policy Act of 1969, the Federal Water Pollution Control Act of 1972, the Marine Protection, Research and Sanctuaries Act of 1972, the Coastal Zone Management Act of 1972, the Ports and Waterways Safety Act of 1972, the Marine Mammal Protection Act of 1972, the Deepwater Ports Act of 1974, the Fishery Conservation and Management Act of 1976, the Clean Water Act of 1977, the National Ocean Pollution Research and Monitoring Act of 1978 , and the Outer Continental Shelf Lands Act Amendments of 1978.

In light of these mandates one can imagine the load placed on the agencies, particularly NOAA, an agency which is less than ten years old. This agency was originally composed of service agencies, whereas the new legislation requires NOAA to be primarily a management and regulatory agency. In any case, the demands on the agency are great, quite recent and growing in complexity. But despite this level of demand, and lack of regulatory experience in some agencies, the "attentive public" and professional analysts continue to claim that what is needed is more policy. I agree with James W. Curlin in his report to the President entitled *The Status of Ocean Policy* that the United States has the most sophisticated policy of all the maritime nations, with over three hundred fifty programs and functions which are administered by ten departments through forty six agencies of the federal government.

I think the criticisms which cite lack of policy are really directed at how policy implementation is organized administratively. This would seem to be the crux of the matter. Many observers find the programs to be "fragmented," lacking in any comprehensive, unified, management concept. These observers yearn for a centralized, independent oceans agency which will direct and coordinate all ocean management, regulation, service and development.

This form of organization places inordinate demands upon the decision-makers. Moreover, a classical model of decision-making most frequently accompanies these organizational arrangements yet—as we shall show in what follows—this decision model fundamentally lacks a capability for adjustment to a complex environment. It also has little capacity to generate the information necessary for effective operation.

In contrast to this model we offer Charles Lindblom's strategic-decision-making approach, which requires a multitude of strategic decision-makers negotiating and bargaining with each other through a process of mutual partisan adjustment.

We shall first discuss and compare the synoptic with the strategic model. We shall then examine the implementation of the Fisheries Conservation and Management Act of 1976 as an interesting example of mutual partisan adjustment in action. We then suggest a comparative research strategy to empirically investigate the operation of the eight Fisheries Management Councils, both with respect to their dynamic interactions with the Secretary of Commerce and the National Marine Fisheries Service. Lastly, we argue for a collaborative cross-national research effort in the Fishery Conservation and Management Act to take advantage of the rich variations in institutions, progress and policies as these relate to marine policy.

Let us now turn to the initial distinction between the synoptic model of decision-making on the one hand and the strategic model of decision-making on the other hand. We shall then turn to an examination of the institutional structure associated with each type.

SYNOPTIC DECISION-MAKING

This is the "classical" approach to decision-making endorsed by a large number of specialists and informed laymen. Risking simplification, the characteristics of this approach to decision-making are the following:

> Faced with a given problem,
>
> A rational man first clarifies his goals, values or objectives and then ranks or otherwise organizes them in his mind.
>
> He then lists all important ways of (policies for) achieving his goals.
>
> And investigates all the important consequences that would follow from each of the alternative policies.
>
> At which point he is in a position to compare consequences of each policy with goals,
>
> And so chooses the policy with consequences most closely matching his goals.[1]

This approach calls for a systematic canvassing of possible alternative policies, for a similarly systematic analysis of the consequences of each possible alternative and for policy choices to serve as goals or objectives somehow separately established.

This approach has enormous appeal, mainly because it requires a single "correct" solution to an identifiable problem. The fundamental flaw in the approach, however, is that it requires what Lindblom calls "a superhuman comprehensiveness."[2]

Despite the fact that most analysts see the synoptic model as only a rough approximation of the world, they nevertheless require well-defined "problems" which call for "solutions" and demand that the policy approach be as "comprehensive" as possible.

Indeed, this is the approach many critics of United States marine policy employ. But, as we shall try to show, this approach is fundamentally defective analytically and therefore critiques based on it are, in turn, weakened. The synoptic approach ignores the fact that information is costly to acquire and the search for it is oftentimes not worth the cost. And since comprehensiveness requires extremely high levels of information, it follows that comprehensiveness is simply too costly. Instead, we must labor to make decisions on policy under the powerful constraint of incomplete information as well as the costs of the analysis itself. In this sense the comprehensiveness required by critics of United States marine policy cannot, and, indeed, should not be maximized.

What models will capture what the decision-maker actually does? Again, Lindblom offers an explanation: the decision-maker adapts–*adaptation is the key to survival in the face of complexity.* Yet it is precisely the synoptic ideal which is not suited to adaptation. The synoptic approach is not adapted to:

1. Man's limited problem solving capability.

2. The inadequacy of information and knowledge.

3. The costliness of analysis.

4. The openness of the systems of variables.

5. The analyst's need for strategic sequences of analytic moves.

6. The diverse forms in which policy problems actually arise—the problem is in fact a cluster of interlocked problems with interdependent solution.[3]

But if one accepts this critique of the synoptic approach, what realities of public policy must be taken into account in order to develop a different, realistic, perspective which permits adaptability along the lines indicated above? In fact, there are empirically observable systems which represent alternatives to the synoptic approach. We shall examine one of these—namely strategic decision-making—in what follows.

STRATEGIC DECISION-MAKING

Charles Lindblom offers an alternative to the synoptic approach which he terms the "strategic" approach. *This approach explicitly accepts the limitations of the human analyst,* and suggests a series of strategies particulary well-suited to these limitations.[4]

Most decisions, he claims, are directed at relatively small or incremental changes based on low levels of information. These are decisions about particular problems rather than a comprehensive view of existing program alternatives. The analysis is also endless and takes the form of an indefinite sequence of policy moves. For Lindblom, policy is *remedial, serial* and *exploratory.*[5]

To pursue incremental changes is to direct policy toward specific ills —the nature of which is continually being reexamined—rather than toward comprehensive reforms. This strategy he terms *disjointed-incrementalism:* the strategy contains the following elements:

> 1. Margin-Dependent Choice: The policy maker is concerned only with the margins at which social states may be changed from the status-quo. He focuses on consideration of policies and then proceeds through comparative analysis of no more than the margins rather than any comprehensive comparison.
>
> 2. Restricted Variety of Policy Alternatives: The analyst deals with a small number and variety rather than the full range of policies that he might be expected to consider. He considers only those which are incrementally relevant.
>
> 3. Restricted Number of Consequences Considered for any Given Policy: By omitting consideration of one set or another of important consequences this greatly simplifies the analysts' task. To omit is to make manageable.
>
> 4. Adjustment of Objectives to Policies Rather Than Policies to Objectives: The ends of public policy are governed by means. One cannot do what one does not have the means to do. Hence, one operates by doing what one can do by looking for objectives which match capabilities rather than vice-versa.
>
> 5. Fragmentation of Analysis and Evaluation: Analysis and evaluation take place at a number of centers rather than one since the problems are too complex to be reached[6] in a single-centered, comprehensive, fashion.[6]

Those who employ the strategy will behave approximately as follows:

The analyst makes an incremental move in the desired direction without taking upon himself the difficulties of finding a solution. He disregards many other possible moves because they are too costly (in time energy and money) to examine. For the move he makes he does not trouble to find out what all its consequences are. If his move fails or is attended by unanticipated adverse consequences he assumes that someone's next move will take care of resulting problems. If his policy making is remedial and serial his assumptions are usually correct[7]

STRATEGIC DECISION-MAKING IN COMPLEX ORGANIZATIONS

To this point we have spoken largely in terms of single decision-makers and have not considered the applicability of the two alternative models to complex, large-scale, public organizations. The organizational setting is critical, however, in that large scale organizations develop patterns and routines of behavior which reinforce incrementalism and serial behavior.

Here the insights of John Steinbruner in *The Cybernetic Theory of Decision-Making* offer a view of organizational behavior which compliments Lindblom's. Both base their arguments on the fundamental assumption of man's incapacity to deal in any comprehensive manner with the complex demands of the environment. Steinbruner, like Lindblom, argues that the key to his cybernetic approach is the *control of uncertainty.* The decision-maker focuses on a few incoming variables which eliminate any serious calculations of probable outcomes.

> "The Cybernetic decision maker is sensitive to information only if it enters through an established highly focused feedback channel and hence many factors which in fact affect outcomes have no effect in his decision process.[8]

Steinbruner also notes that greater complexity entails greater variety, and under conditions of great complexity the decision-maker must have a more elaborate response repertory. Internal simplicity can be preserved under complexity if the number of decision-makers concerned with the problem is increased. Each decision-maker can then focus on some limited dimension of the complex problem. Lindblom has recommended a similar solution, in which coordination is maximized through a multiplicity of strategic problem-solvers which permits *mutual partisan adjustments.*[9]

The latter is contrasted with the synoptic approach to this problem of complexity, which is to array the organization in a single-centered, hierarchical, fashion with information flowing upward to the decision-maker. Unfortunately the evidence is quite substantial that such a design leads to information distortion and greater error proneness.[10]

Given these arguments, and the claimed respective advantages of one model over the other, let us turn to an application of the argument to a particular public policy; namely, the implementation of the Fisheries Conservation and Management Act of 1976.

THE FCMA: A CASE OF MUTUAL PARTISAN ADJUSTMENT

This act has proven to be extremely controversial, mainly—it will be argued here—because it is coordinated through a multiplicity of decision points, and relies on mutual partisan adjustment. Yet this is precisely why we choose to consider it here, in that it would seem to be a particularly promising model of organizational design in the ocean policy area which takes advantage of strategic interactions in order to minimize the inherent limitations of single decision-makers.

The objective of the FCMA is the attainment of "optimum yield." This refers to the amount of fish that will provide the greatest overall benefit to the United States in food production and recreation. It is calculated as *maximum sustainable biological yield from each fishery, as modified by social, political or ecological factors*. Such a system obviously has enormous information requirements for each of these factors and—as we shall see—most of it is quite incomplete.

Under the provisions of the Act, fishery management plans are prepared by Regional Fishery Management Councils, which have representatives from the federal and state government, local communities and private interests. The councils prepare the plan but the plans are implemented by the federal government. In other words, the councils recommend fishery management plans to the Secretary of Commerce, while the Secretary can appoint nominees submitted by the governors together with designated members. To a great extent the councils are a new form of government in that they do not resemble traditional state, local or regional bodies or intergovernmental advisory boards.

The National Marine Fisheries Service (NMFS) plays a support role to the councils by providing operating funds and data on which domestic quota decisions are made. The NMFS also performs a role for the Secretary of Commerce, advising about which of the plans are basically sound. This dual role of serving the councils as well as the Secretary of Commerce is seen by some as a cause of some confusion.

The management technique which is the most controversial is the concept of limited entry. This refers to reducing the amount of fishing effort permitted in a particular fishery. The act permits the councils to recommend schemes for limited entry. Licensing, fees and quotas are the three primary tools used for limited entry. The enforcement of limited entry has been controversial and nowhere has this been more the case than right here in the Northeast. Let us consider some of the properties of the system.

The federal government is charged with implementing the Act in close cooperation with the regional councils. What sorts of information would the Department of Commerce need to implement the Act if it were to act—not as it is currently designed jointly with the regions—but rather in a single-centered fashion, using a synoptic approach in order to obtain optimal yield as it is defined? At least the following information would be necessary:

1. Information on fish stocks;

2. Landing information;

3. Characteristics of the various fishing industries;

4. The political preferences of state and local officials;

5. The social and ecological consequences of any decision.

It goes without saying that no central decision-maker will have complete information of this kind. Yet decisions must be made which involve limited entry. How would this be done by central coordination? The answer is, badly. Information on issues four and five would have to be systemically gathered at great cost rather than be automatically forthcoming, as under the regional fisheries councils, who have partisan reasons for making their views known and also have an interest in specifying unforseen negative consequences. But under central direction, this information would have to be gathered at great cost. And since the conditions and preferences would change, such information would have to be constantly updated. The situation would seem to be a case requiring a disjointed incrementalist approach and not a synoptic approach.

Indeed, the most effective and least costly way to accomplish this task is not to centralize but to decentralize, as in the FCMA, and thereby employ a multiplicity of decision points which, in the process of mutual partisan adjustment, automatically bring forth information that would not otherwise be obtained. This is accomplished in the FCMA via the triadic relationship between the Secretary of Commerce, the Regional Councils (and within the councils) and the National Marine Fisheries Service.

This multiplicity serves to maximize a diversity of viewpoints and preferences. Under such a system there is a higher probability that negative impacts of particular causes of action will be brought to the surface than if the system were centrally coordinated.

This system substitutes social interaction for comprehensive analysis. This interaction is played out as a struggle between competing interests. Conflict is inevitably high in such a system but it may be used to bring out facts which would not be otherwise discovered, given the necessary limitation of analysis. In Lindblom's notion of strategic behavior, *conflict can be used as "a mechanism of discovery."*[11]

Given the fact that information on fish stocks is insufficient and landing information less than complete, precise analysis for optimum yield is impossible. It would seem reasonable then to use interaction and mutual partisan adjustment as a means to discover the preferences of the key actors.

Let us briefly return to the strategy of disjointed-incrementalism. One of the rules for this strategy is that the decision-maker will consider only a limited number of consequences for any given policy. In the case of limited entry it would seem likely that many actors will consider the economic and political side-effects of the decision and be very vocal in making these views known. So, again, multiple decision points will bring out preferences. But beyond this the ecological and social consequences of the policy usually receive short shrift. I suspect that omitting consideration of these factors, as Lindblom points out, allows the decision-makers to concentrate on the other factors in making their decision. Again, to omit is to make manageable!

Another dictum of disjointed-incrementalism is adjustment of objectives to policies rather than vice-versa. NMFS and the Secretary of Commerce simply do not have the means to gather sufficiently reliable data to assure that limited entry decisions are being made properly. Indeed, they must rely on the fishermen themselves for landing data and more recently for information on where the fish were actually caught. They simply cannot obtain highly reliable information in a non-obtrusive manner. Moreover, they are constantly trying to upgrade the quality and quantity of the information. In this regard their behavior is also consistent with the remedial, serial and exploratory pattern of behavior characteristic of disjointed-incrementalism.

Finally, given the differences in the nature of particular fisheries, differences in the characteristics of the fishing industry and differences in the ecological, political and social settings of fishing regions around the United States, it would make sense to proceed by doing analysis and evaluation of each region rather than attempting to use a uniform standard which can be applied in each region. This is the major strength of the institution structure of the FCMA. It permits coordination through a multiplicity of decision points. When there are many decision-makers pursuing problem-solving as a strategy, they strengthen the process of minimizing neglected adverse consequences. As Lindblom argues, "Hence, fragmentation of policy-making to a multiplicity of strategic problem-solvers achieves, we now say, coordination, though, perhaps only in a weak form."[12] The notion that the-more-numerous-the-actors-the-more-coordination seems strongly counterintuitive. It should be the case that the more numerous the actors the more difficult the coordination. But as Lindblom notes, when the decision is incremental and disjointed, more rather than fewer decision-makers can facilitate coordination. Because of the necessity for a division of labor where decision-making is dispersed over a large number of actors in complex situations, their very numbers afford the possibility of coordination.[13]

The precise nature of the bargaining which takes place between the Secretary of Commerce and the regional fisheries councils is not well understood and remains a potentially fertile ground for investigation. Lindblom specifies three different types of bargaining and adjustment which may be present in this process: 1) calculated adjustment; 2) deferential adjustment; and 3) parametric adjustment. In calculated adjustment

> X considers repercussions for Y before making his or her decision, even though he is able to make his decision without Y's cooperation. It is the kind of decision-making in which X's decision is designed to reduce or avoid injury to Y, or to offset injury with benefit so that Y's decision will not be disadvantageous to X.[14]

Deferential adjustment is defined as decision-making intended to avoid adverse consequences for another decision-maker. In parametric adjustment

> the decision-maker simply takes another's decisions as he finds them and adapts them without deferring to or calculating repercussions on others. The adverse consequences of others for him are not forestalled, but they are reduced in effect—at an extreme wiped out—by his autonomous decision to make the best of the situation for himself. X reduces adverse consequences on himself by making a decision designed to cope with adversity to his own advantage.[15]

Which of these varieties applies to the relationships between the primary actors in implementing the FMCA remains to be seen. But it is at least a distinct possibility that different forms of adjustment may be employed by the Secretary of Commerce in relation to different regional councils. A series of hypotheses regarding such interactions and adjustments would seem to be worthy of empirical test. Indeed, the whole system of mutual partisan adjustment which underlies the institutional arrangements of the FCMA should be subjected to careful examination.

INSTITUTIONAL INTERACTIONS UNDER THE FCMA

Our previous discussion suggests that we attempt to obtain empirical data on the observable dimensions of the system underlying the FCMA. In order to do this we argue that an internal comparison be made of the eight US Fisheries Management Councils. Hypotheses will be derived and tested which refer to data on the following features of the Regional Councils:

1. Structure of the Fishing Industry

2. Structure of the Fishery

3. Political culture of the regional councils

4. Nature of bargaining within the council

5. Coalition formation

6. Information levels on the fishery in the region

7. Bargaining relationships of the Council to the Secretary of Commerce (e.g., calculated adjustment, parametric adjustment, or deferential adjustment

8. Relationship of the Council to the NMFS

9. Types of representation on the Council

At a bare minimum, data on each of these dimensions could be gathered and then examined against council management plans and decisions and, in turn, related to the decisions of the Secretary of Commerce. This would yield important data on the dynamics of the councils themselves in relation to characteristics of the fishery, the fishing industry, as well as the interrelationship of these to decisional outputs at the secretarial level. This research should permit us to access the degree to which the FCMA process corresponds to the dynamics associated with mutual partisan adjustment. But more importantly the institutional structure underlying the FCMA may prove worthy of replication in other areas of marine policy. Indeed, these arrangements may prove to be a primary mechanism to avoid some of the inherent limitations of large, complex, government bureaus and agencies. To that extent, we have the potential to more effectively implement marine policy.

CONCLUSION

The search for "comprehensive marine policy" will be fruitless as long as our expectations are influenced by criteria drawn from a synoptic approach. Once we relax the unrealistic demands of such an approach we can see that a decentralized approach which employs a number of agencies at the federal and state level is in no way necessarily inferior. Indeed, it is a mechanism to take advantage of the benefits that the process of mutual partisan adjustment can yield. We have attempted to demonstrate that the institutions which underlie the FCMA exhibit such a capacity for adjustments. But before one can reach any conclusions regarding the respective benefits of these structures, comparative research needs to be undertaken regarding the make-up of the councils, their characteristics in relation to the fishery and industry, as well as their interactions with the Secretary of Commerce. This paper suggests one approach to research on this important question.

But the respective merits of mutual adjustment versus central coordination remain to be seen. These questions can only be answered ultimately by comparing marine policy in the United States with its counterparts in systems which rely more heavily on central coordination. Fortunately, we have knowledgeble observers who can provide such a perspective. It is my sincere hope that comparative marine policy research can be done cross-nationally in a collaborative mode using common designs. I would argue that this is the only way that we can effectively employ our expertise.

NOTES

[1]Charles E. Lindblom, *The Policy Making Process.* New York: The Free Press, 1968, p. 13.

[2]Charles E. Lindblom, *The Intelligence of Democracy: Decision Making Through Mutual Adjustment.* New York: The Free Press, 1965, p. 39.

[3]David Braybrooke and Charles E. Lindblom, *A Strategy of Decision: Policy Evaluation as a Social Process.* New York: The Free Press, 1963, p. 113.

[4]Charles E. Lindblom, *Politics and Markets.* New York: Basic Books, 1977.

[5]Lindblom, *Strategy of Decision*, p. 74.

[6]Lindblom, 92, pp. 94-104.

[7]Ibid., p. 123.

[8]John D. Steinbruner, *The Cybernetic Theory of Decision.* Princeton: Princeton University Press, 1974, p. 66.

[9]Lindblom, *Intelligence of Democracy,* p. 157.

[10]Gordon Tullock, *The Politics of Bureaucracy.* Washington, D.C.: Public Affairs Press, 1965.

[11]Lindblom, *Politics and Markets.*

[12]Lindblom, *Intelligence of Democracy,* p. 155.

[13]Ibid, p. 157.

[14]Ibid.

[15]Ibid., p. 158.

CHAPTER 18

Public Policy For A Specialized Interest: The Oceans

BRIAN ROTHSCHILD and JUDITH ROALES

National Oceanic and Atmospheric Administration
United States

INTRODUCTION

Ours is an age of slogans. We all remember: "The New Deal;" "This is our finest hour;" and "Tippicanoe and Tyler too."

The Ocean community has its slogans, too! Recall now "The Stratton Commission;" "Will the outcome of UNLOS be successful?" "Oceans reorganization;" and "Yes, we have no ocean policy."

It is important for us to consider how these slogans fit into the fabric of our day-to-day thinking and activities. To what extent have these slogans served as mere justifications for ongoing activities and the study of non-problems or, on the other hand, to what extent do they provide a "light at the end of the tunnel," channeling our energies and resources toward some useful purpose?

In order to explore this question and evaluate the appropriate direction for new policy initiatives and research, let us examine the new crop of ocean slogans. One in particular appears to be dominant in our jargon—it is "ocean management."

We need to ask: "To what extent will 'ocean management' be a justification and stapling together of ongoing activities or, on the other hand, to what extent will the term 'ocean management' rally our resources and energy to produce an overall betterment of society?" As a result of interest in "ocean management," will we in fact have new ways of addressing societal problems and new ways of mobilizing our talents and resources to aid in their solution? What will change as a result of "ocean management," and will the costs associated with managing the oceans be worth their expenditure?

Before we pass on to make judgments regarding ocean management and new directions in ocean policy implicit in the use of the term, it would be well to indicate what we mean by the term "management" in the first place and how we judge its worth.

WHAT IS MANAGEMENT?

Perhaps the simplest way to think of "management" is in terms of a shoe factory. A shoe factory, for example, has certain inputs: leather, heels, soles, shoelaces, labor. It also has certain outputs: a variety of different kinds and sizes of shoes. The activity of management involves making the myriad of decisions that result in producing the different kinds of shoes from the variety of inputs. The quality of this management is measured by the profitability of the shoe factory operation.

By analogy, ocean management starts out with inputs of the oceanic resources: the minerals, the fish, the coastlines, the waste-sink capacity of the ocean. The ocean managers then make decisions or non-decisions regarding the use or non-use of the resources. These decisions result in the benefits or "disbenefits" which society receives from the resources. The quality of ocean management is measured by the extent of benefits received by society compared to the "maximum benefits" possible.

In order to judge our performance in ocean management, we have to understand what we mean by "maximum benefits." In a simplistic, abstract, theoretical, quantifiable world, the level of maximum benefits is often obvious. The limited entry concept in fisheries is a perfect example. Many of the early limited-entry thinkers believed that the appropriate amount of fishing effort was that which maximized each fisherman's profit. Most fishermen and anybody else would agree that this is a good criterion reflecting maximum benefits—to the fishermen—as a result of fishery management. But fisheries rarely operate at a level where this theoretical result is even close to possible. Because, *interalia,* the abundance of fish is uncertain and different from year to year, attainment of maximum theoretical benefits would require corresponding changes in the number of fishermen. When stock abundance was low, it would be necessary to prohibit some fishermen from pursuing their livelihood. This, of course, is generally politically unacceptable in the US.

The point of discussion is not to discourse on limited entry, but to demonstrate that theoretical or, better yet, idealized notions of maximum benefits are seldom attainable in the real world because the collection of decisions which must be made, that is the management process, is fraught with uncertainties, unknowns, and external considerations. They are:

Factors that are intrinsic to the activity being managed (e.g., the daily production of fish or the cost of drilling per barrel of offshore oil produced);

Factors that are extrinsic to the activity being managed (e.g., the political situation in the Mideast);

Decision chains that are long and complex (nearly one year to deal, procedurally, with a fishery management plan);

Inappropriate models of the way the world works (e.g., Keynesian economics or a deterministic baseline view of the environment); and

An imperfect and untimely flow of information to decision makers.

We can see then that the uncertainties and unknowns caused by these factors are important forces governing the performance of a management system. This points out that in some cases apparently poor management results are actually the best that can be derived from the management system, given the decision environment in which it must operate. The only way to improve the performance of such a system is to identify the uncertainties and unknowns so that they can be dealt with in an appropriate and explicit way.

Any success that ocean management has will not relate specifically to the breadth of its authority or to the efficiency with which the day-to-day programs are run. Rather, it will relate to the degree to which the activity of ocean management reduces the uncertainties and unknowns that are so dominant in ocean-resource utilization.

Many of the uncertainties and unknowns in oceans activities relate to the fact that ocean affairs is not a dominant societal concern. Thus, ocean managers need to put their interests in the context of national and indeed global issues. In addition, they must appreciate the intensity of societal anxiety and the frustration of solution which is so prevalent.

Society in the US now functions in a climate of serious and overriding concern for our economy and increasing energy shortages. The news is rife with slogans of anxiety: “double-digit inflation;” “combatting OPEC;” the need for “an energy doctrine;” “balance of payments;” “synthetic fuel;” and “declining productivity,” just to name a few.

These slogans are but the crust on the onion soup. Deeper in the bowl we have crises in poverty, transportation, nutrition, pollution, health care, and recreation. Whether these concerns are intellectualized or are, in fact, real, they dominate the anxieties of nearly every American—to the extent that little public anxiety is left over for the oceans.

The impact of these problems is compounded by frustration with attempts to devise solutions. Such frustration is evident in the hallowed halls of the intellectual community. Bernard Nossiter (*Washington Post,* May 20, 1974: B1) reviewed the state of the braintrust. He quoted Nathan Glazer:

> At cocktail parties, people say, "Isn't Carter terrible?" He replies, "Why I don't know what I would have done. The cupboard of ideas is bare. If Carter had task forces, I wouldn't want to read their papers. Now there is nothing. No new departures."

> ...There is in fact an "end of concensus with respect to what to do. (It)...has driven many younger, bright academics into a retreat from public policy. Ambitious economists elaborate elegant mathematical solutions to theoretical problems with little, if any, relevance to public issues." Sociologists and political scientists devote themselves to testing and evaluating programs and scrupulously avoid a search for fresh answers.

Nossiter hit the nail on the head when he added:

> The great intellectual breakthroughs rarely precede but are typically the product of events. If answers to problems are found, they are more likely to emerge from experience, from trial and error, than from some startling departures in the learned quarterlies or in academia.

UNCERTAINTY IN THE DECISION PROCESS ITSELF

Another aspect of uncertainty and unknowns is implicit in the very decision processes themselves. These decision processes often had their beginning in state and federal legislatures. They now exist in a new environment where freedom of resource-use conflicts with new conditions of real or apparent resource scarcity; where short-term concerns and anxieties outweigh longer term and perhaps more important concerns; where accelerated mobility and communications have heightened the notion of sovereignty to the extent that the common good is sacrificed for the good of individuals and groups of individuals to a greater extent than ever; where in many cases the management of the oceans with their unique and complex properties is viewed in the same context as the management of land; and where our only forum for having an international dialogue on ocean issues, the Law-of-the-Sea, is fading into the past.

The decision processes relating to the oceans are not special or unique. The basic policy directions which motivate most of our public decision processes are the laws of our land. And we often find laws that have intrinsic inadequacies in that they are technically incorrect or they generate unworkable procedures. We find laws that are in conflict with one another, and we find laws that are difficult to interpret or administer. In addition, there are various societal problems that are denied possible solutions because political standoffs prevent legislative consideration of the problems.

All of this is not surprising when one considers the increased technical complexity of the problems confronting Congress; the sometimes spotty quality of the research that Congress and the executive branch depend upon for an information base, and the fact that the incentives for legislators to act on behalf of the nation are not nearly as great as the incentives for legislators to act on behalf of their districts or special interests.

The organic function of the legislative branch is to exert a fundamental control on the capability, the legitimacy, and the tenor with which the executive branch administers the will of Congress. It is well known that nearly every major federal agency can find in its charter some mission that gives it access to the ocean; at least twelve cabinet-level departments or agencies are deeply involved in some way with ocean activities. A listing of the legislative authorities found in public laws which deal directly with the oceans runs to nearly four hundred pages. Hundreds of city, county, state, and regional bodies can all, in a variety of ways, choose to control or ignore the coastal ocean and thus impact upon the territorial sea. In general, the involved agencies lack interest in and an understanding of the synergism and antagonism of multiple uses and the capacity of the ocean to withstand uses. At all levels of government, in many cases the management goals of the bodies are as temporal as party politics; in most cases the management schemes are products of the expediency of allowing existing systems to be manipulated to solve immediate crises which, more often than not, are political in nature. In fact, the resolution of multiple short-term crises in such a manner often leads to inadequate long-term solutions. The result of much of the combined activity of the legislative and executive branches is an increase in regulating the activities of the citizenry in spite of considerable concern with respect to the efficiency of regulatory processes. In fact, the President through a recent executive order has directed his agencies to make the regulation process more efficient and responsive to societal needs.

A recent analysis by John Quarles (*Washington Post,* May 20, 1979, B8) points out, however, that it is impossible to streamline or eliminate many regulations because they are mandated by Congress. The nature of the legislation itself guarantees a poor performance.

Quarles says, "It is often argued that regulatory agencies are overzealous. Sometimes that is true, but more often agencies take the heat from Congress. The implications of statutes are seldom widely understood at the time of enactment. Later, when an agency implements burdensome requirements, a typical reaction is to blame it for decisions which had been made by a congressional committee." He goes on to observe that time frames contemplated in legislation are often unrealistic; that detailed requirements in the laws on occasion do not make sense when applied nationally, and that the increase in regulations (*Federal Register:* 13,500 pages in 1974 and 27,000 pages in 1978) has exceeded the capability of agencies to enforce; that legislation is often cast to bring pressure on various groups or to advertise congressional policy and as a result it is often not as efficient as it might be for administering the law. Quarles says:

> Congress must mend its ways. It must insist on better analyses of costs and benefits and administrative workability—before it legislates. It must exercise a new restraint in controlling the length and specificity of legislation.

Given the existing body of legislation, the executive is faced with a considerable number of laws that are equivocal and ambiguous to execute. This organizational nightmare makes it difficult to focus accountability, to learn where decisions are in fact made and for what reasons. In fact, many decisions never really get made. Many important problems evidently become lighter than air and float above the agencies themselves into the twilight of the interagency process.

Interagency mechanisms come in an assortment of shapes and sizes, with varying degrees of formal structure and a range of duties. Although some interagency relations are mandated by law, that is most often not the case. The relations grow from some common objective and a need—perceived by someone somewhere in the system—to facilitate cross-communication. The official, given purposes of interagency groups are generally to reduce the conflicts and costs or to capitalize on the benefits of our pluralistic federal programs. The groups bring together opposing programs or actions in order to effect a compromise, or they bring together similar and related programs in order to strengthen a product. The unofficial purpose, or hidden agenda, is generally to legitimize agency and individual activities, to co-opt other players in order to minimize criticism, or to give the appearance of action while other kinds of less-official decision-making processes are operating.

Since 1966—when the Congress passed the Marine Resources and Engineering Development Act which resulted in the Stratton Commission and the Marine Science Council—interagency panels or groups have been very much in style. As the government role has become ever more complex, interagency relationships have undergone exponential growth.

The relationships range from two-agency agreements for a specific project to multi-agency agreements intended to coordinate policies and programs.

When the General Accounting Office made its 1975 report, "The Need for a National Ocean Program and Plan," it counted up fifty major multi-agency committees. In the intervening years, some of these groups have disappeared, but others have been instituted to take their place. The number of major, formal committees appears to have remained about the same. But the number of two-or-three-agency committees or working groups has become almost limitless.

For example, at the end of 1978, to the best of its knowledge, the National Oceanic and Atmospheric Administration was party to more than 125 Interagency Agreements, more than 330 Memoranda of Understanding or Agreement with other agencies, and held more than 125 positions on interagency committees or panels.

And these 600-plus interagency relationships do not include all of the "reimbursable" agreements—that is, one-time contractual arrangements in which one agency pays another for services or personnel.

Note that this number—six hundred—is the count "to the best of

NOAA's knowledge." It, no doubt, includes duplications and defunct committees because it is difficult to get a true picture of interagency activities. In NOAA and most other agencies there is a specified clearance procedure for reimbursable contracts, as well as internal regulations regarding who can sign such contracts. The larger the dollar amount, the higher the level of the signer. However, this is not always true for interagency agreements which involve program coordination or policy questions. There is no official definition of who may enter into agreements, no set clearance procedures, no guidelines for the structure and function of such groups. There is only a directive which requests that such agreements be reported to a central file. Compliance with even this request is not always monitored; the files are not kept current.

NOAA is not alone in its confusion about interagency relationships. In attempting to assess interagency relationships late last year, we contacted the Department of the Interior, the Coast Guard, the Environmental Protection Agency, and the Navy. Interior could supply no listing of its interagency activities. The Coast Guard supplied a list of nearly seventy agreements last updated in March 1976. EPA had a detailed computerized list of contractual agreements, but no count of policy or program agreements. The Navy could supply a partial list of about fifty unclassified agreements of various kinds and indicated they were in the process of computerizing their system for better tracking. None of these agencies could supply lists of the membership positions they held.

The purpose of mentioning the current chaos in interagency relationships is to demonstrate that this too is a decision environment which is difficult to predict.

Interagency groups with specific programmatic purposes and relatively exclusive membership tend to be successful at the working level. However, as the scope of the relationships becomes more theoretical and more people are involved, success becomes more elusive. At the far end of the spectrum, where many agencies are gathered to work out crucial overarching policies, success is nearly non-existent.

Numerous groups have reported consistently for more than a decade that the Federal Government needs a national focus for its ocean programs, that there is increasing need for coordination of activities across agency or department lines, that there is a lack of progress due to the fragmented nature of authority and the scattered funding. Yet the interagency groups have been unable to change the situation. Their charters are written with slogan words like "evaluate," "facilitate," "coordinate," "identify," "foster," "encourage," "review." Regardless of how well-intentioned they may be, they have no power to implement. Their problems are very basic:

> They have no direct authority, that is, no budget to implement their recommendations and no budgetary control over the programs about which they make recommendations.
>
> No one person is really in charge.

> Although representation generally begins at a high policy level, participation rapidly slips down to a level where the attendee cannot speak for the agency or commit its resources and personnel.
>
> They make decisions by consensus and compromise. The final product must please all members; therefore, it rarely has specific, tough recommendations. The language becomes general enough that all members can continue to protect their turf and their budgets in spite of the committee's work to coordinate.
>
> If there is unsolvable conflict, the process sooner or later degenerates into a complicated method of passing the problem out of the committee and agency members and up to the Office of Management and Budget for resolution. This puts the decision-making process into the hands of those who are several layers removed from any active expertise on the subject in question.

In the rare instances where there has been modest success through use of an interagency committee process, it has been achieved by totally voluntary cooperation, or because members perceived that there was a real power behind the committee. This was the case, for example, with the Marine Science Council which was chaired by Vice President Hubert Humphrey. Successes also generally come in growth situations, where all members stand to share in a gain, such as a new program, a budget increase, or a surge of public interest.

If the interagency mechanism is to work in the future to help resolve conflicts of ocean use, it must do more than provide a forum. It must become a decision-making process with more accountability. The number of organizational entities in any interagency relationship contributes to a situation in which the members become advocates for the area of their responsibility, rather than advocates for the public good. Advocacy in the agencies breeds standoffs and lack of decision. As in the legislature, the executive branch harbors members with conflicting direction, minimum accountability, and an orientation toward keeping an activity going, rather than providing specific output for the public.

Without clearly defined measures of performance oversight, effective management and control are exceedingly difficult, not to mention the stifling effect of the unavailability of relevant and timely information.

All of this may sound like a diatribe against government. It isn't. It is simply a description of many of the general—though important—problems with which an ocean manager must contend.

We have now created, albeit in a very cursory way, a societal context for ocean management; that is, the process of making the myriad of decisions that effect the oceans. By way of quick review, we have set up a framework which acknowledges that: (1) ocean management is made up of the set of decisions that society makes regarding the use of ocean resources; (2) the quality of these decisions (e.g., the benefits that accrue to society as the result of their effort) depends on the degree of uncertainty and ignorance in the decision process; (3) this uncertainty and ignorance can be reduced by expanding the scope of the activity to address these factors explicitly; (4) the anxieties of the man in the street are substantial, and they do not relate in any significant manner to those of the ocean community (with the possible exception of some aspects of the energy problems); (5) there is considerable frustration in government and the intelligentsia regarding just what to do about the malaise that is so prevalent in our contemporaneous outlook; and (6) the legislative-executive system is often shackled by cumbersome problems and an unfavorable decision environment.

We think that any ocean management policy needs to explicitly take into account the contextual environment that we have described. If it does not, then the range of benefits that accrue from said ocean management will be limited, narrow, and stilted.

But what should we do? The possibility of changing our system of operation is certainly beyond the horizon.

Obviously we must lower our sights and search for new policy direction, prognostications of the success and failure of which can come by a careful examination of comparative policy.

RESEARCH

What policy advice and concentration can be given to an agency whose competence covers a specialized interest such as the oceans? We haven't caught up with idealistic policy goals—we need an alternative to "looking ahead." The following could be considered:

1. Develop organization patterns that enable agencies with non-dominant policy interests to effectively interface with the dominant agency. For example, an ocean agency in these contemporary times would need to have clusters of expertise on the economy and on energy.

2. Put weight on dealing with direct resource conflicts by generating and disseminating information. For example, the catch of anchovy in California is thought to be at a low level because the recreational fish require the anchovy for food. This is not in fact known. If the facts were known, then this resource conflict could be minimized.

3. Minimize the chances of Congress and the Administration making technical error by focusing analyses of cost-benefits and administrative feasibility on the legislative agenda.

4. Develop new guidelines for accountability in the interagency process. Perhaps every interagency committee should have a specific decision agenda and an independent arbitration mechanism.

5. Make organizational capability and management efficiency a matter of agency policy.

6. Develop an informed public. It is important to have people who have the intellectual basis to deal with policy issues.

7. Develop a global perspective in the agency. Global events—in terms of

commodities, the environment, and the economy. Drive national and local policy issues. Develop regional competence as well as federal competence in this area.

8. Develop an approach to change views of sovereignty with different degrees of resource use.

9. Bridge the gap between academia and the forums where policy is actually made.

10. Develop better cooperation, for example, between the buyers in a lease sale and those others who deal with the problems.

11. Take a holistic, systems approach. Avoid disjointed incrementalism.

Much of what we have said is well known or slightly innovative but not often acknowledged or, where appropriate, implemented. We believe that the policy direction needs to inquire into the future of implementation. The critical need in the oceans is not to gain new authorities or to establish even more articulate policy goals, but rather to be able to use what we already have to do a better job for society. If we can begin to do these things, then "ocean management" can be a meaningful concept, as well as a slogan.

Commentaries

ROGER BENJAMIN

University of Minnesota
United States

We need anticipatory policy-relevant designs in the comparative marine policy field because the pace of change itself outdistances our capacity to react or respond effectively to the results of actions already taken or events that have occurred. Thus it is the case that we know—and especially in the case of major policy areas such as the oceans—that we must anticipate the direction as well as the impact of future as well as present trends if we are to mitigate the worst case scenarios which abound. Professor Wenk's chapter has developed an elegant set of policy goals for which anticipatory designs must be built. The problem is that the assumptions, the images, about development underlying Professor Wenk's critique of the present and his design suggestions for the future are built on questionable assumptions about development and planning. Here I shall sketch out some implications of recent work in comparative political change and political economy that provide us with a different perspective from which to construct a set of assumptions and images about development more appropriate for design work in the comparative marine policy field. Let me attempt to justify these assertions by starting with some points of agreement with the provocative points made by Professor Wenk (this commentary should be read in concert with his chapter.)

Professor Wenk is surely correct in his critique of the current incoherent state of national and international marine policy. Moreover, his bleak assessment of the negative ecological consequences of current trends, for example, pollution, overfishing, seabed exploration, and the rest, indicates the irreversible consequences that soon will be upon us. Professor Wenk also notes the artificiality of the distinction between marine policy and public policy, that is, between national marine policy and national and international public policy. And this is an important point because it means, indeed, that the field of marine policy is to be best regarded as a subfield of public policy. I have argued elsewhere that the most difficult problem facing students of public policy is the absence of structuring principles with which the central problems of the field may be defined and ordered and connections with other approaches made. If this point has merit, the assumptions and images underlying whatever passes for present theory or modeling efforts in a subject area—such as the one under examination—require searching examination. For if the assumption concerning the importance of the pace of social-economic and political change is valid, the way we think about and conceptualize the present trend serves to determine what we see as the future. At the risk of excessive simplification, I turn to summarize the assumptions about development underlying Wenk's chapter to note his prescriptions, and to conclude with an alternative view.

Underlying the Wenk critique is the fear that the future will contain much of the past and the present; the assumption is that the present trends will simply continue into the future. What are the basic properties of the present? First, that the global community of nations may be divided into the standard first-second-,and third-world categories. Second, economic and especially technological development drives social and political change. Moreover, there is the largely unquestioned assumption that the technological change will continue to be linear in nature—onward and upward. This is not a welcome assumption, of course. Professor Wenk, for example, laments the lack of long-range planning in political institutions, the absence of leadership, the absence of national—let alone international—coordination of marine policy. The Wenk prescriptions argue for some sort of centralized set of institutions that can solve these difficulties. In other words, to handle the problems—presumably abetted by fragmented political institutions—planning and coordination of policy solutions must be generated in large-scale, hierarchically organized institutions. Here, however, I wish to sketch a different perspective concerning the direction of solutions to the basic problems in the comparative marine policy field. And, in order to proceed, I shall take a short detour through recent work in comparative political change and collective goods theory. The intent is to suggest a different set of assumptions regarding development and the centralization solutions offered.

For observers in the nations represented at the Comparative Marine Policy Conference (Western Europe, Canada and the United States, plus Japan), there is growing realization that "development," as traditionally defined no longer captures the change taking place in these countries. This appears to bode both good and ill for the solutions envisaged in the "ten commandments" outlined in the Wenk chapter. Foremost is the decline in economic growth as understood through the economic vocabulary describing market activity. In all of the nations noted above, the service—including the public—sector outstrips the industrial sector in terms of growth rates. Moreover, there are related social and economic changes that add up to a fundamental transformation, a transformation from the industrial era that emphasizes growth to the postindustrial period that emphasizes quality of life issues. The changes include an emergence of postbourgeois values (an emphasis on nonmaterial wants), better educated public, population growth slowdown, a decline in citizen trust of political authority, decline of political parties, and the rise of political demands on government, especially from local action groups. Indeed, in our postindustrial societies, the age of politics appears to be dawning. However, if the arguments concerning the slowdown of growth have merit, man may again have the necessary time to redevelop the bonds of community destroyed during industrialization.

Why is this political transformation underway? Why are the liberal-democratic regimes that exist in our group of countries under assault? The argument, based on collective goods theory, runs as follows. First, one of the qualities of postindustrialization concerns the growth of interdependence in all areas of life. In the words of a Bell telephone commercial "the system is the solution." On the negative side, this interdependence also produces a crowding effect; more citizens find it necessary to protect their privacy and autonomy. Increased interdependence also leads to an increase in externalities (spillover effects), that is, unintended side consequences of the production and/or consumption of goods. These externalities may be positive—or they may be negative, for example, as in the case of air and water pollution. Second, I have noted the increase in educational levels of citizens and a basic change in many of their preference schedules from an emphasis on material to social-psychological wants. These two points have much to do with our political predictions because of the following argument. Goods may be divided into public, private, and collective (or mixed, quasi-public, quasi-private) categories. "Pure" private goods are the familiar divisible goods which are bought and sold in the market place. Externalities presumably do not enter in because the pricing mechanism is wedded to supply and demand and this results, over time, in the achievement of an equilibrium. Public goods are indivisible; their production results in all citizens in the community (catchment) to which they are delivered receiving the same amount of the good. Typical examples include national defense, fire and police protection. Our collective goods cover those goods that externalities adhere to. They are goods about which there is confusion over who is to receive what proportion and who is to bear what proportion of the costs. Moreover, these are the goods, such as public power plants, regulation from the public sector, subsidies, and private transportation, steel and oil companies, that produce negative externalities. We may, in the absence of excessive overall tax burdens, not care about the free rider problem in the public sector, but the presence of collective goods *inevitably* drives citizens into collective action units. In postindustrial societies collective goods become increasingly predominant. Finally, a point about the relationship between institutional size and efficiency and the delivery of public and collective goods. From standard public administration theory, the dominant models in comparative politics, it is largely received wisdom that with greater size comes greater efficiency, that is, economies of scale. The benefits of centralization presumably include greater capacity to plan because there is room for more professionals with the relevant critical skills useful for planning. And coordination of formerly fragmented activities may be developed. However, if we assume the distinctions about goods noted above, it is reasonable to question the centralization argument. For collective goods, especially, larger organizations may only exacerbate the problems which create political conflict. Because of the information-sensitive nature of collective goods, their delivery may 1) require smaller rather than larger government institutions to deliver them; and 2) lead to inevitable disputes concerning collective goods, which may best be accommodated in institutional arrangements that promote bargaining, trading, and log rolling.

What I have sketched comprises elements of an argument that leads one to question the linear extrapolations about change made by most observers and participants here. No matter whether one views this argument with alarm, as some do, or with the optimistic sense that such change represents a significant opportunity for man to redesign his political institutions to more closely approximate classic democratic theory ideals, is immaterial. If the argument has merit it puts discussion about marine policy questions in a new light. For example, discussion about marine policy within our postindustrial societies will be more—not less—complex in the future as economic and ecological tradeoffs are grappled with. However, it is in the comparative, international dimensions of marine policy that the implications of this alternative-development model of postindustrial societies and the collective goods language used to interpret the development pattern are of special interest.

Let us recall the description of the problem facing comparative marine planners in the Wenk chapter; that it is an accurate assessment there can be little doubt. The problem is that the nations not represented here, from the second and third worlds, are at points in the development process where they are willing to bear the costs of the negative externalities so well summarized by Wenk's chapter. They are willing—again in the language of economics—to internalize these externalities because their priorities remain material in orientation. Unless substantial incentives or sanctions are provided for these countries to cease their roles in destroying the oceans, eloquent statements of the Wenk variety will be regarded as liberal cant and ignored. Second, if the centralized political institutions of our postindustrial countries do continue to weaken, there will be declining leverage points to move the oceans from their present position as an example of the tragedy of the commons. And, finally, if we break down the goods provided by the oceans, some like pollution do affect countries differentially; pollution, until its advanced stages, is a collective, not public evil. Military problems, fishing, minerals, also affect different parts of the globe differently. Centralized planning units are unlikely to be the appropriate institutional design to handle the above problems. Rather, we should seek to work out more flexible institutional arrangements that deal with functional areas of marine policy (see the Hennessey chapter).

In sum, at a time when calls for centralized, coordinated comparative marine policy are being made, the centralized national planning structures in postindustrial societies are in disarray. To move toward a centralized institutional design for the oceans would simply create another disjuncture between individuals and groups and the problems they must deal with.

LENNART LUNDQUIST

Uppsala University
Sweden

The preceeding discussion shows that there are a number of problems, there are a number of standpoints, there are a lot of arguments. Everybody is calling for a comparative marine policy to solve the problems, to sort out the arguments, to find out whether or not standpoints are valid. As far as I understand, there is no such thing as a field of comparative marine policy today. There is no common terminology. There is no common view as to what questions should be addressed. But if there were such a field, in what way should it be organized so that it would become relevant to the problem-solving that everybody is crying for? That is what was just referred to as "the basic structuring question," which is something I will try to look at a little bit.

Let me start out by saying that I am a political scientist and I look at comparative marine policy from the angle of the political scientist. There is one thing I don't want comparative marine policy to be, and that is the study of comparative government, because as I will try to show you in a minute, organizational structure and institutions for allocating and exercising responsibility are only one part of the marine-policy problem.

What I would like comparative marine policy to be all about is the *policy* part of the problem, the content of the marine-policy goals, actions and arguments hopefully forthcoming from governmental institutions. But even if we agree to concentrate on the *policy* content, there are so many different arguments within the discipline of public policy studies, public choice, whatever you call it, that I just simply don't find that we could at once be relevant in the sense that practitioners would understand what we are saying. There is this gap between policy scientists on the one hand and policy practitioners on the other. As policy scientists, we'll have to clear out a lot of terms, concepts, and arguments before we can present the practitioners with a systematic, disciplined, and relevant comparative marine-policy science.

So I think maybe I could use my reactions to Timothy Hennessey's chapter for just going ahead with some of these suggestions. Hennessey's chapter concerns the organization of part of the US fishing policy. His chapter tries to lead us to believe that these councils are the most efficient organization to achieve the purpose of US policy. They are less expensive in terms of decision-making costs and other organizational patterns.

The Councils involve regulated interests. This might make the decisions more easy to implement. You have—as the Swedish expression of this hostage principle goes—"anchored your decisions before they are going to be implemented." Hennessey's chapter indicates the types of organizing framework that I would like to see us using in comparative marine policy studies. It is policy-oriented in that it takes as its point of departure the practical *problem*, in this case the distribution of responsibility. It tries to drive home a *standpoint*: the councils are *the* most efficient type of organization for this job. The *arguments* used in the paper are founded on scientific research on complex organizations and on public choice.

In my way of thinking as it has developed over a couple of years, scientific research as well as practical political debate boils down to three things: *problems, standpoints* and *arguments.* It is surprising—and to some extent very depressing—to find that so much scientific research lacks this simple organization. Think for yourself! How many papers have you read where there was no problem addressed? There was a fine statistical job with nice correlations, but what was explained, what kind of thesis was supported? What hypothesis did they falsify or verify?

Comparative marine policy to me concerns six problems, no more, no less. The classification is not arbitrary; you get it by simultaneously adopting two perspectives.

First, there are two ways of looking at marine resources and the marine environment. The first way is to look at it in terms of *maintenance* and *reproduction.* Then you are concerned with the problem of keeping marine resources at a level providing a reasonable yield, reasonable in terms of what levels of welfare, what objectives we want to achieve. The second perspective is one of *utilization* and *consumption* which is concerned with getting out all of the amount of resources necessary—at the moment—to fulfill our welfare needs. Whether you use the utilization-consumption way of looking at it or whether you use the maintenance-reproduction way of looking at it, you find there are three different types of values to be allocated in the marine policy.

The first is *welfare*; some of us call it benefits: How much of what marine resources are to be conserved or utilized, and who should receive the benefits?

The second value to be allocated is *responsibility*. How should the rights and duties associated with marine resources and the marine environment be distributed among the *levels of government*, between the *public* and *private sectors*, and among *individuals* and *groups of indivduals?*

The third value to be allocated is *costs*: How should the sacrifices that we have to make in order to get the marine benefits and to have the marine policy organization be distributed over *time*, among *individuals* and *groups* in society? Thus, we have a five-fold matrix: *maintenance-conservation, utilization-consumption,* and then we have *welfare, responsibility* and *costs*. Thus, we have five scientific and practical issues to discuss in comparative marine policy.

The thought behind this five-fold matrix is that no one of the five problems is more important than the other. A reasonable strategy for comparative marine policy studies will be to systematically vary the central research problem. If responsibility for conservation is the central research problem, standpoints on the other problems count as arguments in the debate over what the allocation of conservation responsibilities should be. They become possibilities and constraints on the problem of conservation organization.

In all these five boxes, comparative marine policy studies could proceed by using social and legal scientific methods, as well as relying on results from other sciences, to ask questions ranging from policy initiation to policy impact. How is the situation, what does it look like? *Why* is it that, for example, certain countries have a certain kind of organization and not another? How *effective* is the organization? How *effective* are the means that are used? Why do certain means work in a certain context but not in others? What impacts seem to be related to what policy means?

If we think of these five problems as five central problems in comparative policy research, we should not forget at any time that we also want it to be relevant to practitioners. We have to remember that the decision-making perspective is central. There are so many comparative policy studies that try to find out whether there are statistical correlations between, for example, the socio-economic background variables on the one hand and spending patterns and policies on the other. And it's simply a sort of explanatory endeavor. You never know in what way it would be relevant to policy-makers, because it turns out that the most important thing in policy-making, namely, the politics itself, is never accounted for. You just account for the importance of socioeconomic variables and you never come down to things that are in fact much more easily changed or manipulated by political situations.

If we want to become relevant to marine policy practitioners, we must thus remember that policy problems are problems of *decision-making*. First, the decision-maker has to set objectives: what is desirable? Second, he must measure the problem scope, as I call it, to find out what is lacking in the present situation: how large is the gap between goal and reality? Third, he must select means: what should and could be done in marine policies? Finally, there is the problem of policy evaluation: how effective are the means? This is one of the central things that I want to bring out here. We should move much more in the direction of policy evaluation studies. Means of a similar type or dissimilar type are used to solve the same problems. How effective are the means, and what impacts are there? Are the impacts solely dependent on the fact that a particular kind of means is used? Or are there other particular circumstances and conditions at work which are specific to one nation or which are at work in other nations, too? In other words, can we isolate the effects of different means from the institutional background and from the effects of other background factors?

To conclude, if comparative marine-policy studies are aware of what different standpoints and arguments mean in the marine policy debate, they could indeed contribute a lot to practical policy-making by critical research into the empirical validity of the political argument. The fact that we have heard a lot of arguments, or that this or that policy alternative is impossible to use because of this or that particular constraint, means comparative marine-policy studies could assess the validity of these arguments. It might turn out that they are just myths or simplifications. What comparative marine-policy studies could and should find out is whether national marine policies—and the arguments used to support them—do indeed stand up, given the empirically verifiable results and experiences of marine policy measures in other nations. In that way, we could establish a field of research that is both scientifically disciplined and politically relevant. That we owe to man and sea alike.

DOUGLAS JOHNSTON

Dalhousie University
Canada

I previously thought there was probably a considerable overlap between law of the sea studies, which have developed impressively over the last 15 years, and the emerging field of comparative marine–policy studies. I thought that those who have studied ocean affairs from an international perspective in the last 15 years would find ways of adjusting to the comparative perspective in the next five or ten years. I am not so sure any more.

There are, however, some resonances to be detected, even admitting that the field of comparative marine–policy studies is not likely to have at its center anything comparable to UNCLOS III. The impact this conference has had on the volume and nature of social science research is incredible, and I doubt that anything remotely comparable to it in size and complexity is going to occupy center stage in the field of comparative studies. But then debates arise in the national context from time to time on major legislation being put forward, such as the recent fishery management and coastal zone management legislation enactments in the United States. Generally, these debates don't go on (thank goodness) for ten or fifteen years, like the law of the sea debate, but they may last long enough to pull academics into various roles that may make them something other than wholly detached and dispassionate.

There are really three questions, it seems to me, that have to be raised, if not answered, at this stage of anticipation. First, how will researchers in this comparative field perceive their roles in relation to the actual process of marine policy–making and ocean management? Even admitting that it is not an entirely identical setting, we have learned something surely from the last fifteen years in the law of the sea literature. Between the two extremes—the highly constrained employee doing research under maximum constraints, dictated by the terms of his service on the one side, the (probably mythical) totally detached academic observer, on the other—there have emerged in the law of the sea field many types of roles that are neither one extreme nor the other, but a bit of both.

If we move along the continuum from the totally constrained employee doing research and scholarship, probably the first variant that we come across is that of the academic who is drawn in under a contract as adviser or consultant to a government or to an industry, or some other kind of organization with a very strong institutional interest in the matter. This person is under some constraint by reason of confidentiality and discretion, but not to the same extent as the employee. Further along the continuum, we come to what I would call the partisan researcher, who is not an employee of an interest group, but has a sectional interest in what is going on. He identifies very clearly, sometimes very emotionally, with a particular set of issues that belong to a larger context. He is a kind of advocate. We know many people of this kind in law of the sea, and I would suggest we will meet many scholars of this kind also in the comparative field.

Next we come to the type of researcher I would call a structural researcher, who structures his research in a particular way, idealistically or ideologically, that colors his whole approach to his research. It may even be deep intellectual commitment of a particular type of framework in inquiry, a methodology. Whatever the reason, his structural approach gives him focus, but it may also limit him more than other types of researchers. He is in the middle of the continuum, between the totally detached and the totally attached. He is "semi-detached," if I can borrow from real estate terminology.

Then, moving more towards the detached side, we reach the researcher who sees himself as correcting deficiencies. These are perceived deficiencies that may take the form of omissions or underemphasis or overemphasis. He sees himself as having an obligation to perform to the intellectual community, as it were, but not as a representative of any institution. Last on my list, close to the detached observer, is the researcher who thinks of himself as monitoring what is going on, not from any narrow perspective or sectional point of view, but in a manner suggestive of concern that something is going wrong, or at least is likely to do so if it is not closely watched. No doubt you could add other types, but it seems to me the ones I have mentioned, perhaps others, will be discoverable in the field of comparative marine- policy studies.

The role that the scholar will adopt in the comparative field will also be determined by the strategy that he consciously adopts for himself. I am not sure how many strategies one would want to set out, but two at least come immediately to mind. One strategy would be designed so to ensure that the researcher's approaches to the topic chosen, the set of tools employed, the techniques applied, the methodology pursued, the data base exploited, should be compatible. The "official" or wholly "attached" researcher is unlikely to have an incentive to challenge the accepted techniques, the existing data base, and so forth. He will presumably come from time to time to different conclusions, albeit using the same data, the same techniques but under the strategy based on compatibility there is an inherent limitation on his capacity to furnish society with an alternative judgment. He is more likely to reinforce previous judgments and opinions than to arrive at an entirely different, alternative view of things. For fairly obvious reasons of convenience and advantage post-academic researchers will tend to align themselves with the strategy of compatibility.

The other model is that of the researcher who thinks of himself as required to provide a balance. It is precisely because of the "official" research that precedes him that an academic specialist may adopt a corrective strategy. In official circles a researcher may have little choice but to pursue the strategy of compatibility, but different assumptions and self-perceptions come down from the tradition of academic inquiry. The truth is, of course, that the academic community today is highly heterogeneous, and a wide range of views exist on the role of the academic in society.

The second question that I wanted to raise was this: What range and pattern of linkage arrangements should we expect to emerge between university research programs in the field of marine policy studies? This question assumes the relevance of infrastructure in a field such as comparative marine-policy studies, which by its very nature must be cross-disciplinary, large-scale, and world-wide. It's a hybrid field of inquiry, even if it is done solely within a single national setting. We are talking about major programs and projects which can be supported by only a few universities in the world, which have the right mix of capabilities to compete effectively with the best-endowed government research departments. Universities tend to think in competitive rather than cooperative terms about one another, but somehow they will have to overcome the academic yearning for autonomy. Academic scholars have not shown a notable willingness to come forward and share, but I suspect they will do so in comparative marine-policy studies, because they are likely to perceive it as necessary and not as a matter of choice.

Let me name some universities which are involved in the large-scale development of the field of marine policy studies. The University of Washington in Seattle has focused on Northwest America and the Pacific rim; the East-West Center in Honolulu on the South China Sea, and perhaps later on the South Pacific; the University of Southern California on the islands of the South Pacific. In France, at Nantes, they are beginning to focus on the Northeast Atlantic Ocean. At Dalhousie (if I may add my own university), we will be focusing on the Caribbean and West Africa, as well as Atlantic Canada. There may eventually be a dozen or so universities with large-scale research programs in the field, studying almost every marine region in the world. Those now coming forward with proposals to foundations for initiating other major programs will want to think about the geographical focus that they should apply to their program, since no program can focus equally clearly on all parts of the world together. Those already involved realize the complementary nature of this kind of research, and I think the newcomers coming into the field will wish, as we have done at Dalhousie, to build into our budget allocations for effecting inter-institutional linkage arrangements.

What do I mean by linkage arrangements? Well, probably an exchange of researchers for limited periods of time, probably exchanges of trainees or research assistants jointly sponsored and financed. Joint conferences are already familiar to all of us, and even joint projects. It is not necessary that a consortium of universities need be effected. Most kinds of linkages can perhaps be executed on a more or less informal basis, at least in the immediate future of the 1980s.

The third question we have to know how to answer, before anticipating the shape of things to come, is this: To what extent will the academic researcher be able to secure access to those research tools that are most appropriate to the research he wishes to undertake? Social scientists are notoriously ineffective, I believe, when it comes to the choosing of research tools. They generally allow themselves to be victimized by dogma. Lawyers are the worst offenders of all, I hasten to say. We are so much less sophisticated in this matter than our colleagues in the physical sciences.

Our failings and shortcomings in the sophisticated choice (and use) of research tools may be a serious problem we have to learn to overcome. Let me illustrate the point by referring to the computer. How many social scientists make frequent and really sophisticated use of the computer? Now it might be said that our students know more about computers than we do. New types of computers are coming "onstream" all the time, in increasing numbers, especially in the more affluent countries. In trying to project the future of comparative marine-policy studies around the world, we should recognize that computer-assisted research, intelligently conceived, is bound to be of benefit almost anywhere, at least to accelerate the early phases of marine policy studies. But until there has been a sufficient transfer of skills, the advent of advanced computers will increase the gap between the research that can be done on the campus of a rich country and what can be done on the campus of a poor country, and add asymmetry to the field of comparative marine policy studies.

Related to the question of computers is the question of the bibliography. What role should the bibliography now play in the world of the comparative marine policy analyst? I think we have to shed some of our older ideas about bibliographies. If it is sheer mass and volume of titles that one wants to secure, one need only retrieve from the data bank, but arguably need not publish in print form.

The future need in a field such as ours might be for specialized bibliographies, and annotated bibliographies above all, to save time. It can be a very time-consuming business to provide an annotated bibliography, but these are some of the questions that we have to be prepared to address in future conferences such as this.

BARRY BUZAN

University of Warwick
United Kingdom

I am going to make a few comments on Timothy Hennessey's chapter, and then will go on to make a few comments about new directions in comparative marine policy as they might relate to international studies, which is what I am most interested in.

Hennessey makes a distinction between two kinds of procedural approaches to decision making, and these are labeled "synoptic" and "strategic." The synoptic approach is characterised as being centralized, hierarchical, expensive, likely to distort, and not representative of what happens in the real world. This seems to be a reasonable critique. Offset on the other side is the so-called strategic approach.

This seems to be a terrible misnomer. What we have is a fragmented, adjusted kind of process. How that name was determined is a mystery to me.

This is supposed to work by substituting, as it says, social interaction among many competitive actors for comprehensive central analysis. This certainly reflects reality in some respects. It doesn't cost as much and it is argued that this procedure will achieve a measure of coördination arising from integrated interest.

My sense of the chapter was that Hennessey supported this latter approach not only because it reflected what was happening, but also because it was somehow a better procedural approach than the other. I would like to take issue with this contention.

I think there is a critical statement in the chapter which goes a long way towards tempering any enthusiasm one might have for this strategic approach. He claims that "ecological and social consequences of the policy receive short shrift versus the economic and public interests of the participants." It seems to me that we are precisely concerned about the lack of accounting for social consequences of policy. Therefore, if one adopts a procedure like this strategic procedure instead of the synoptic procedure, then it is those ecological and social consequences which are going to receive short shrift, and that is what people are upset about.

It is, to my mind, dangerous to characterize this as a procedural issue. It is fundamentally a political issue. Looking at the outcomes and also looking at the practices of these kinds of approaches, if one takes a fragmented, incremental, marginal-adjusting strategy, that is not only a way of proceeding, it is a basic political choice. It is very much the kind of basic political choice that reflects the American political system, as opposed to the more centralized European systems. It seems to me that what this "strategic" approach does—to put it in generalized terms—is to generate the typically American outcome of private affluence and public squalor.

The European system, it might be argued, produces the reverse: private squalor and public affluence! You do have to make a very fundamental political choice. It is dangerous to treat these as administrative alternatives. What is at stake here are basic political questions. That is my reaction to the chapter. I think it raises a very important point because one needs to consider—as indeed both chapters here have done—what the implementation strategies are and what the implementation procedures for these policies are.

There must be a lot of interesting middle ground between the extremes of the synoptic and strategic approaches. One can find ground where one is not stuck with the choice of two extremes.

Let me go on in a rather different tack to add something general about new directions in comparative marine policy studies. It seems to me that any direction is going to be a new direction, given the almost complete lack of comparative studies at the moment. There is a great shortage of comparable national studies of policy making, or even of policy, in this field as in many other fields, and such studies as exist are mostly in-house government projects. Many of the chapters here reflect this: a lot of these chapters are of the kind that one finds circulating within government bureaucracies for in-house use. But these tend not to get out, so people outside the government don't necessarily know what the thinking on these issues is, and people outside the country certainly don't know what the thinking is.

These matters have not been widely circulated, and in that sense there is no literature of any great extent on comparative marine policy studies.

I think, however, that there is quite a good case for producing such literature. Being that we don't have the literature, what is the justification for producing it? As I said before, since my own interest is in studies of international ocean policies, I am going to take that angle and see what it is about a potential literature in comparative marine policy study that might help out on the international plane.

Robert Friedheim's chapter can provide the analytical basis for advocacy in a more rational international environment. I want to elaborate on this a little bit. I think there are some fairly common errors that arise out of that area, some of which have been illustrated here.

One of the common errors is the assumption that the policy of others is coherent and rational whereas you policy is constrained, confused, fragmented, and so forth. This we have heard a lot. I think this is a false perspective, or at least not nearly as true as it first may appear. If you come from a European country and you look at the national policy problems of the European country through a perspective within that country, it certainly doesn't look coherent, rational and well-planned. It looks confused, pluralistic and all of the other adjectives that have been brought up to describe American policy.

If from an American perspective you see European policy as rational and coherent, you are assuming certain things about the way we behave, and those assumptions are wrong. When you encounter us in the negotiating forum, you assume that we are being coherent and rational and don't have all the difficulties that you have, but we in fact are looking at you and feeling you are coherent and rational and we are burdened with domestic troubles, and why are you making so much difficulty! It seems to me that understanding this could remove a lot of unnecessary obstructions to negotiation.

Another angle to this is the assumption that the objectives of other actors are the same as your objectives. If you assume they see the problem as you do, then they must have the same kind of objectives as you do. This is not true a lot of the time. Americans in international negotiations are often concerned with policies that can roughly be labeled "efficient" according to certain sets of values, highly held in the United States, whereas other actors, developing countries in particular, are more concerned with something called "equity," which isn't efficient. If you judge the policies and positions of other states according to the values of your own policy, then they often seem irrational or hostile, whereas they are not, they are just different. That needs to be understood, and better policy studies showing what it is that other states are concerned about would help. Not just what they are concerned with, but why and how they are concerned, in a way that will enable people to talk to each other more clearly and concisely.

That is quite a strong justification for more working in comparative marine policy studies. There are some other aspects worth mentioning—that is, if you accept my argument so far, and assume that almost everybody sees himself in terms of his own policymaking as being slow, incoherent, irrational, constrained and pluralistic, and that this is a major reason why international negotiations are so difficult and so slow.

If you go to the Law of the Sea conference, one of the problems is that many of the states there—including the United States and many of the European states—still haven't sorted out their own policies. If the United States hasn't sorted out its policy, it is very difficult for anybody else to do anything because they can't negotiate with somebody who doesn't know what his position is. This leads to interesting questions which can be explored by comparative marine policy studies.

Another question relates to the domestic decision-making structure which both these chapters addressed. Are these domestic structures outdated? Are they simply hopelessly old and old-fashioned and not really useful for the kind of policy problems that they are facing, or are they merely inappropriate to deal with an environment in which things have become more international? Are they actually quite good in terms of meeting domestic needs, but not very good when they have to interoperate those domestic needs with international ones? This would make an interesting set of questions to ask. There is clearly a dilemma over the collapse of the boundaries that define foreign policy. This lapse of boundaries is easily illustrated by the European problem of whether or not European Economic Community matters are foreign policy or domestic policy. That is an extraordinarily difficult question to answer, and the kinds of machineries needed to cope with this are not yet well understood, and might be an object of study. There is a very clear need for comparative marine policy studies simply to encourage cross-fertilization of ideas. That has happened a lot here, and has been very interesting, so that simply the comparing of policies will provide more options for solutions to problems which everybody has.

If you know what other people are doing, that might give you some idea of what you can do. A particular item worth concentrating on is the case for a unified Ministry or Department of Marine Affairs versus the highly fragmented decision-making and administrative structures that exist in most places now. There is strong lobbying in many European countries for this kind of department. This would go very much against the line of thinking in Hennessey's chapter, but could be considered in abstract. Is this an appropriate area for unified policy-making?

Finally, to end on the note on which I began, I think another justification for comparative marine policy studies is to keep the soon-to-be unemployed army of Law of the Sea Conference anaylsts in employment—in other words, to provide a new pond for old ducks!